Age-Dating Stars

Age-Dating Stars
From the Sun to Distant Galaxies

Maurizio Salaris

CRC Press
Taylor & Francis Group
Boca Raton London New York

CRC Press is an imprint of the
Taylor & Francis Group, an **informa** business

First edition published 2021
by CRC Press
6000 Broken Sound Parkway NW, Suite 300, Boca Raton, FL 33487-2742

and by CRC Press
2 Park Square, Milton Park, Abingdon, Oxon, OX14 4RN

© 2021 Taylor & Francis Group, LLC

CRC Press is an imprint of Taylor & Francis Group, LLC

Library of Congress Cataloging-in-Publication Data

ISBN: 978-0-367-44399-3 (hbk)
ISBN: 978-0-367-44124-1 (pbk)
ISBN: 978-1-003-00946-7 (ebk)

Typeset in Computer Modern font
by KnowledgeWorks Global Ltd.

To Adam

If thou hast knowledge, let others light their candle at thine
(Thomas Fuller, 1727)

Contents

Preface

Early in the morning of 2 February 2001, I took a taxi from home for the hour-long trip to Manchester Airport, in the North-West of England. I had a very long journey ahead of me. From Manchester I flew to Heathrow Airport in London, followed by a second long (and very uncomfortable) flight to San Francisco, and a third one from San Francisco to Honolulu, Hawaii. My plan was to spend the night in Honolulu before the next day final flight to Hilo, the oldest city in the Hawaiian archipelago.

I arrived exhausted in Honolulu late in the evening of the same day (thanks to the time difference with England), but I had made the capital mistake of not booking in advance a room for the night, convinced that it was easy to find a place in one of the many hotels near the airport. I had no idea that Honolulu was hosting the American Football NFL Pro-Bowl the next day. Obviously, no free rooms in town, and I had to spend a sleepless night on an airport bench, reading an atrocious science fiction book I had bought in Manchester before starting the long trip, and making friends with the airport cleaning staff busy working through the night. I finally arrived at my hotel in Hilo around lunchtime on Saturday, destroyed by fatigue. The night between Saturday and Sunday I slept about 12 hours, which allowed me to recover and absorb the jet-lag, ready for Monday.

Was this 'odyssey' worth it? Granted, it was not quite like Dave Bowman's transcendent trip in '2001 A Space Odyssey' that led to his transformation into a star-child, but I definitely did not regret it. On the following Monday, the conference *Astrophysical Ages and Time Scales* started, five exciting days of talks, posters, and discussions to address in an astrophysical setting the following questions: 'When did this event happen, or started to happen?', and 'How long did it take?'. These are two of the most fundamental questions to be answered in everyday life, be it a murder investigation, the study of archaeological finds or biological processes, or to understand why your car or your fridge has stopped working.

In the case of astrophysics research, it is the knowledge of ages that tells us how far in the past ('when') an event happened or started to happen – the beginning of the expansion of the universe or the formation of the Sun or of our galaxy, for example – and how long it took to reach completion. For this reason, measuring ages and in particular the ages of stars, is one of the traditional goals of astrophysics. Stars provide us with a network of timepieces distributed across galaxies and the whole universe, and thanks to the development of the theory of stellar evolution, we can 'read' the time marked by these clocks, and use this information to help address major astrophysics questions.

The idea for this book was triggered not just by the importance of the ages of stars, but especially by the rather surprising – at least for me – realization that the age of celestial objects (including the Sun) and the age of the universe are often an antagonizing subject in the public domain.

In May 2017, the American Academy, in partnership with the Carnegie Institution for Science, hosted a meeting titled 'Communicating Science in an Age of Disbelief in Experts'. A report about the meeting makes clear that one of the reasons for this disbelief is the lack of explanations of how the scientific process works. Science is not just an accumulation of facts, it is important to explain how we get evidence and draw conclusions from the evidence. I wondered whether at least a partial reason for the rejection of the ages of stars and the universe established by us astrophysicists comes from a lack of explanations of how we really do it. In the process, I came to realize that it is very hard to find accessible and at the same time comprehensive descriptions of how we determine stellar ages.

The purpose of this book is to explain how we can read the times marked by these cosmic clocks, including a historical perspective, a self-contained description of the various steps that have led us to develop the techniques we use to establish the ages of the Sun, nearby stars and stars in distant galaxies, what broader questions the ages of stars help us to address, and some main results. I have presented the methods that more directly give us stellar ages, by comparing observations of some properties like for example radius, brightness or chemical abundances, with theory, that I would call primary methods. Some other techniques exist that, however, make use of the ages of stars established with these primary methods as calibrators.

I have taken special care in explaining what we do and why, by following (hopefully) clear and logical steps, with focus on the main concepts rather than the details. Along the way, the reader will find a concise summary of our current view of the universe and its historical development, the theory of stellar evolution that underpins almost all methods to derive stellar ages, and crucial tests (also little-known ones) of its validity. I also didn't shy away from discussing uncertainties and open problems. An appendix contains additional information about a few selected topics that could not be expanded in the main sections of the book without breaking the logical flow of the exposition.

This is not meant to be a standard academic textbook, for it is aimed at a much broader audience. The necessary physics and astrophysics background is built up in the various chapters and in the appendix, and I strived to keep the level of the presentation as non-mathematical as possible, apart from very few simple formulae here and there, that give a more quantitative feeling of the concepts I am discussing. The book contains the set of differential equations we solve to calculate models of stars; however, their purpose is just to show the translation to the mathematical language of our description of how stars work, to predict the structure of a star and its evolution. I think it is very important to make the reader aware of this essential step, but I certainly keep away from discussing how we solve these equations. There are some technical descriptions in a few chapters that are unavoidable if I wish to explain what we really do to determine stellar ages. I am confident they can be followed without too much effort.

Each chapter also includes a glossary with some of the most relevant technical terms, and a list of recommended books and papers (both at academic and popular science level) to expand upon the main topics discussed. The bibliography contains a set of references to technical papers for the more advanced readers, that include the sources of the data shown in some of the figures, selected major works, and more technical details on some of the topics discussed in the various chapters.

As I already wrote, this is not really meant to be a classical textbook, but I believe it can be used with profit as an auxiliary text in courses on stellar astrophysics and galaxies at both undergraduate and graduate level, and by high-school teachers to add some up-to-date inspirational research elements in their astronomy lectures.

I am indebted to David Hyder, Andrea Miglio and Alessandro Savino for preparing some of the figures for this book. Paola Cerri, David Hyder and Nate Bastian are warmly acknowledged for reading and commenting on selected chapters. I also wish to thank all my astrophysics colleagues and friends worldwide with whom I have worked (and I am still working, they didn't get too fed up with me yet) during these last 30 years.

Finally, it would have been an enormously harder task to write this book without the NASA's Astrophysics Data System Bibliographic Services, the arXiv e-print repository, Google books, and Project Gutenberg, to which I am very grateful for the services they provide.

CHAPTER 1

You Must Be Crazy

Not long ago, some acquaintances who live in Rome became aware of my job as an astrophysicist in the UK. When they learned the name of the neighbourhood where I grew up until the age of 15, they were somewhat surprised: Science is not one of the expected career paths for my generation raised in the 1970s in the borough of Garbatella.

There are probably few boroughs in the world that know the exact date of their foundation, and Garbatella, in Rome, is one of them. Its birthday is recorded on a plaque stating that on 18 February 1920, in the presence of the King of Italy Vittorio Emanuele III, the first stone was laid of a new neighbourhood: the Garbatella. According to common wisdom, this name comes from the moniker for the owner of a local tavern, not far from the Basilica of Saint Paul. She was a beautiful and graceful lady named Carlotta, so well-liked by the travellers who visited her tavern, that it was referred to as the inn of the lady 'Garbata and Bella' (Polite and Pretty) subsequently syncopated in 'Garbatella'. The reason for the name also goes back to her charitable attitude towards the needy, and today there is a stucco depicting this lady on the facade of a building in the neighbourhood, not far from where my maternal grandparents lived. The name Garbatella also seems to have crossed borders, because when I visited Lebanon years ago to give a seminar at the American University of Beirut, I found a Lebanese restaurant called Garbatella during a trip to the nearby mountains.

The architecture followed initially the English model of the garden cities built for workers that included significant green spaces to provide residents with locally sourced food. After the first small villas came two- and three-floor buildings, then with the advent of fascism a few years later, houses more similar to modern condominiums began to be built.

Figure 1.1 A street of the neighbourhood of Garbatella in Rome, to-day. Not much has changed since its foundation about a century ago (ValerioMei/Shutterstock.com).

They hosted the displaced people who had been driven away from the city centre to make room for Mussolini's public works[1]. Some of the buildings were considered great examples of modern architecture to be shown to important guests visiting the city. On 12 December 1931 Garbatella was even visited by Mahatma Gandhi.

Even today the neighbourhood has preserved its winding streets, steep stairways, and charming little squares. The shade that prevails is red; you can find it in the plaster of the buildings and the exposed bricks, in contrast with the white of the columns, frames, windows, the many decorative elements on the facades, and the black wrought iron of the railings and gates. The other predominant colour is the green of the lush vegetation (see Fig. 1.1). It is the ideal place for a walk among its gardens, and historical places such as the Palladium, a former cinema now used as a theatre[2].

It looks like an idyllic place, but it certainly wasn't easy to live in during the 1970s and the 1980s. Economic crisis, political violence and

---

[1] This is when my maternal grandparents moved to this neighbourhood from their flats in the historical central parts of the city.

[2] I remember watching at this cinema, for the first time, the masterpiece *2001 A Space Odissey*, around 1979−1980.

terrorism, common crime, the unstoppable rise of drug addiction, all had a dramatic impact on life in the neighbourhood. I remember well the owner of the fruit and vegetable shop near our flat, murdered in front of the store because of gambling debts: My mother used to shop there almost every day, taking along my brother and myself during the long school summer holidays. Or the kid who attended my school and died crashing with a stolen scooter. I still have vivid memories also of pavements and parks strewn with syringes, some with blood still fresh on the needle. Not touching syringes and accepting nothing from strangers has been a common thread throughout my childhood. The life of too many people my age who grew up in that neighbourhood ended tragically because of drug addiction.

It was the first months of 1977, and I was in bed with the usual winter flu. The year before, during a monetary crisis, the national currency at the time – the Italian Lira – was devalued by 12%, terrorist violence continued to claim victims, and a disastrous earthquake in the north-east of the Country caused the death of over a thousand people.

At the time I attended the *Pablo Picasso* middle school (a print of the paint *Guernica* was hanging right beyond the school entrance), a three-year bridge between primary school and high school, hosted in a building that no longer exists. I was used to walking there with three or four friends from primary school years. The meeting point was near my parents' flat, and we walked to school as a group, at a brisk pace, going through some narrow streets flanking an old and abandoned warehouse.

At school, I found the lessons generally easy to follow, the teachers were good, and I especially liked maths and science. But bullying and intimidation were standard fare, and it wasn't fashionable to be a good student with top marks. In fact, I will never forget one of my classmates, a girl of whom I still remember the name, telling me one day: *Salaris, good students are out of fashion these days!*

Close to school, there was a bookshop with always a selection of science books on its shelves. I stopped every day to look at the titles and covers, fascinated especially by astronomy books, which were often displayed alongside publications on UFOs and aliens. I loved science books, an interest probably born during visits to the nearby flat of my uncle. The bookshelves in his living room were filled with not only comics and science fiction books but also several encyclopaedias, and during every visit, I spent hours reading about physics, astronomy, and maths in those volumes. Watching live on television the Apollo astronauts walking or driving on the surface of the Moon must have also played a role. I could not know it at the time, but today it looks like an extraordinary privilege to have grown up at the only time (so far) in history when humans set foot on an alien world. Sometimes I wonder whether it will ever be repeated.

A few days before I was in bed with the flu, I saw on display in that shop a book called *Stars and galaxies*, and I had enough money with me to buy it (it's still somewhere at my mother's place). It was a hardcover little book, with lots of colour figures and photos, written by a Spanish professor and translated to Italian. I had the pleasure of meeting the author, Ramon Canal of the University of Barcelona, almost 20 years later, when I was a postdoc at the Institute of Space Studies of Catalonia. However, at the time of this meeting, I did not recall he was the author of that 'fateful' book.

That winter morning, warm under the covers, I began to read about stars and galaxies. It wasn't the first popular astronomy book I read, and I did not expect any earth-shattering new revelation. After a first half devoted to galaxies, in the following sections about stars I discovered for the first time 'stellar evolution'. The realization that stars evolve changing size, brightness, temperature, their inner structure and chemical composition was like an epiphany. First of all, the concept that stars evolve and change was fascinating and made them much more interesting in the eyes of a young boy. Secondly, their evolution coherently explained the various stellar classifications found in the popular books I read before.

Two important points escaped me at the time (after all, I was only 11 years old and not a child genius). The first one is that the Big-Bang (the 'birth' of the universe, we'll see more about this in chapter 2) produces essentially only hydrogen (H) and helium (He) but if you look at the chemical composition of the human body, you'll find that about 65% of the mass of a person is made of oxygen (O), about 10% of H, 18% of carbon (C), 3.3% of nitrogen (N), 1.5% of calcium (Ca), 1% of phosphorous (P), plus small percentages of other elements like magnesium (Mg), potassium (K), sulphur (S), sodium (Na). Apart from the 10% of hydrogen, the other elements must have been produced after the Big-Bang, and indeed the theory of stellar evolution shows that they are all synthesized at high temperature in the interior of stars, and eventually ejected in the interstellar space. Out of interstellar gas enriched by stellar ejecta new generations of stars (and planets) form and evolve, and the cycle is repeated. We owe our existence not only to the light coming from the Sun but also to unknown stars that, before the Sun was born, have produced the elements that make up our bodies. It sounds melodramatic and not very original to state that we are all 'stars' children', but indeed we are. And this is a very sobering fact.

The second implication that I failed to grasp, was that if stars evolve, and if we can mathematically predict their evolution, then it must be possible – at least to some degree – to determine their age. And measuring the age of stars ended up – mostly by chance – to be one of the main lines of my research, after I was lucky enough to satisfy what became my crazy childhood dream of becoming a scientist.

1.1 WHY DO AGES OF STARS MATTER?

Stars cannot talk to us, they cannot show us their identity cards, and evolve on timescales much longer than human (and whole civilizations) lifetimes. Historically, their unchanging appearance in contrast to the terrestrial realm was taken as a sign that the heavens had a fundamentally different nature. Determining the age of stars is a bit like the work of a forensic anthropologist, although not remotely as macabre. Forensic specialists apply anthropology to criminal investigations and help identify human remains, estimating the age of the victims, as we have seen hundreds of times in television series. Their work rests on the knowledge of how human bodies develop over time and is based, among others, on the degree and location of bone growth. Our bones are mostly soft cartilage at birth, which is then slowly replaced by hard bone at over 300 different centres of growth. These centres of growth eventually fuse to form the bones we find in adult bodies. And after the body stops growing, some bones begin to fuse together. Given that we know the speed at which these processes advance, anthropologists can use the observed patterns of bone development to estimate ages.

Along the same lines, if we can calculate reliable models of stars and their evolution, we should be able to find time-varying features to be observed and exploited for age determinations. But why should we be so keen to determine the ages of stars, apart from our innate compulsion to learn about the world around us?

Let's start with an example closely connected to our existence as a species, If we are completely ignorant about the structure and evolution in time of our star, we can't predict if/when its luminosity and its size will change, and by how much. We might make the instinctive and reasonable assumption that nothing is going to change for a very long time, maybe for eternity, but there is no certainty that this would be the case. With the knowledge of the full evolutionary cycle of the Sun, and its actual age, we might instead be able to predict changes that would impact our planet, and (hopefully) prepare to mitigate the consequences.

In a purely astrophysical context, the age of the oldest stars sets a lower limit to the age of the universe (they cannot be older than the universe they live in), and serves as a consistency check for both our models which describe the evolution of the universe ('cosmological models') and the physics we use to describe the structure and evolution of the stars ('stellar physics'). Even if the age of the universe is settled (more on this shortly), the ages of stars are still crucial to study the formation and evolution of our Milky Way and galaxies in general. Age-dating stars allows us to measure the timescales of the processes that drive the formation and evolution of the galaxies; and a knowledge of timescales helps identifying what these processes are.

As an example, let's assume that in one case we find all stars in our galaxy to have the same age, equal to the age of the universe, whilst in another scenario we find a large age range amongst the Milky Way stars, spanning several billions of years. These two scenarios clearly suggest different mechanisms for the formation of the Milky Way, with different processes at work in shaping our galaxy. In the first hypothesis, we are led to conclude that galaxies were already fully formed the moment the universe came into existence, questioning the accepted physical conditions at the Big-Bang. In the second scenario, the formation of the Milky Way has followed a much slower build-up. The age differences, coupled to measurements of the chemical composition of the stars, to correlations with positions and eventually motions in space, give us precious information about what physics has driven the formation of the Milky Way. Stars are indeed like keepers of the past evolution of the cosmos.

Finally, in an era when discoveries of extrasolar planets are coming thick and fast, the determination of the ages of planet-hosting stars is crucial to study the formation and evolution of their planetary systems, and the differences with the history of our solar system. And if (or perhaps just 'when') we will have identified signatures of life from observations of distant planets, the ages of the host stars will be crucial to study the development of extraterrestrial life forms.

Before starting our journey to understand how we determine the age of stars, I wish to discuss briefly two specific examples of scientific debates from the past (recent and not so recent) that show the role played by the stellar ages in their resolution.

1.2   AGE OF THE SUN, GEOLOGY, AND BIOLOGY

Planets form out of the gas and dust left over from the formation of the host star; hence their age must be essentially the same as the age of the star they orbit. This general idea that a star and its own planets form from the condensation of gas and dust (a 'nebula') is around since the early eighteenth century, when it was first proposed for the Solar System by Emanuel Swedenborg, and slightly later in more details by Immanuel Kant and Pierre-Simon Laplace, who argued that nebulae collapse into a star and the remaining material settles in a flat disc out of which planets form. Today observations with radiotelescopes (that can detect electromagnetic radiation at long wavelengths, on the order of millimetres and longer, appropriate to 'see' cold objects) can probe these protoplanetary discs around newly formed stars, and capture the increasing brightness and temperature of the planets in formation.

The age of the Sun has implications for understanding the processes that have shaped Earth and the development of life. The reason is that geological and biological evolutionary timescales must be consistent with the age of the Sun. For a well-documented example, let's shoot back in time to the nineteenth century. In the first edition of his seminal book *On the Origin of Species by Means of Natural Selection, or the Preservation of Favoured Races in the Struggle for Life* (the full title of what is usually called *On the Origin of the Species*, published on 24 November 1859 in London by John Murray), Charles Darwin made a crude calculation of the age of the Earth by estimating how long it would take erosion occurring at the current observed rate to wash away the Weald, a great valley that stretches between the North and South Downs across the south of England. An 'old' Earth is a crucial prerequisite for his theory of natural selection that needs very long timescales to work.

At that time most people in the Western world believed that the age of Earth was about 6,000 years, as estimated in 1644 by John Lightfoot, Chancellor of the University of Cambridge, and in 1650 by James Ussher, the Archbishop of Armagh[3]. However, geologists were starting to gather evidence pointing to a much older planet, and Darwin was

---

[3]Lightfoot published his calculation in the book *The Harmony of the Four Evangelists: Among Themselves and With the Old Testament*, while Ussher presented

heavily influenced by the idea of 'uniformitarianism', championed by the Scottish geologist Charles Lyell, and presented in his three-volume *Principles of geology*, published between 1830 and 1833. The key idea of Lyell was that the present is the key to the past, and that geological remains from the past can be explained by geological processes now in operation and therefore directly observable. Seen through the lens of uniformitarianism, our planet must have existed for an immense period of time, for the measured rates of geologic change are too slow to create the modern shape of the Earth's surface without millions and millions of years of activity.

At page 287 of the first edition of *On the origin of the species*, Darwin wrote: *If, then, we knew the rate at which the sea commonly wears away a line of cliff of any given height, we could measure the time requisite to have denuded the Weald. This, of course, cannot be done; but we may, in order to form some crude notion on the subject, assume that the sea would eat into cliffs 500 feet in height at the rate of one inch in a century. This will at first appear much too small an allowance; but it is the same as if we were to assume a cliff one yard in height to be eaten back along a whole line of coast at the rate of one yard in nearly every twenty-two years. (...) Hence, under ordinary circumstances, I conclude that for a cliff 500 feet in height, a denudation of one inch per century for the whole length would be an ample allowance. At this rate, on the above data, the denudation of the Weald must have required 306,662,400 years; or say three hundred million years.*

This age of about 300 million years sets a lower limit to the age of the Earth. If correct, the age of the Sun should be of at least 300 million years, and its total lifetime longer than that. What was the age of the Sun estimated at that time? How did scientists calculate this age at a time when stars[4] were to a large extent still mysterious objects?

Crucial steps towards addressing this question were made first in 1842–1843, when Julius Mayer and James Joule discovered that heat is a form of energy. In 1850 Rudolf Clausius stated the basic idea of the second law of thermodynamics: Heat cannot spontaneously flow from cold regions to hot regions without an external work being performed on the system. This is why a refrigerator needs to be powered

---

his result in *Annales Veteris Testamenti, a prima mundi origine deducti* (Latin for *Annals of the Old Testament, Traced Back to the Origin of the World*).

[4]Stars were already considered to be celestial bodies of the same nature as the Sun. They were known to be immensely further away than the furthest planet; hence they were shining by their own light, like the Sun.

to maintain a constant temperature lower than the environment. What happens spontaneously is that heat (energy being transferred – see the appendix) flows from hot to cold regions instead.

It became clear that the Sun is hot, much hotter than the surrounding space, and shines because of the loss of energy (in the form of electromagnetic waves, or 'photons' – see the following chapters) due to its high temperature. This means that the solar lifetime can be determined by calculating the ratio between the total amount of energy available, and the energy lost from its surface per unit time, the solar luminosity. The solar luminosity, assumed to be constant in time, was known from measurements of the radiation received on Earth, and the knowledge of the Earth-Sun distance.

In 1854 the physician and physicist Hermann von Helmholtz argued that chemical reactions cannot be the source of solar energy. For example, if the Sun is a furnace burning coal (and the whole Sun is made of coal) its lifetime would be of just about 5,000 years. At a time when nuclear physics was unknown, the only viable source for the solar energy was gravity, more specifically the gravitational potential energy, that we often denote with V. To make a classic practical example, let's consider a book placed on top of a desk. If the book was originally on the floor, some external agent must have worked against gravity to raise it to the desk. For example, I had to pick it up and spend energy to set it on the desk. But if the book then falls back to the floor because I didn't place it correctly, the energy acquired when falling (called kinetic energy from the Greek word kinētikos, meaning 'of motion', because it is associated with the motion of the book) that is transformed into heat and sound upon impact with the floor, is now provided by the gravitational force. When the book is on the floor, we say that it has less gravitational potential energy than when it is on the desk, and the (kinetic) energy acquired during the fall from the desk is equal to the difference in gravitational potential energy between the two locations.

In brief, the gravitational potential energy – I will call it just gravitational energy from now on – is the energy an object has because of its position in a gravitational field. It can be in principle positive or negative depending on where we set the point where $V=0$, but the choice of this zero point is somewhat arbitrary, for what matters are just differences in V. If the floor is the zero of the potential energy, then a book on the desk has a positive value of V, whilst if the top of the desk is the zero point, the same object on the floor has a negative amount of potential energy.

It seemed natural to set V=0 when the particles are an infinite distance away, because the strength of the force of gravity between two particles goes as the inverse of the square of their distance (it gets weaker when particles are further away from each other) and approaches zero only for very long distances. With this choice of the zero point, if m is the mass of a particle in the gravitational field of a mass M, its gravitational energy is always negative, and equal to $V = -GMm/r$, where r is the distance between the two masses and G the 'gravitational constant', a fixed number equal to 6.67×10^{-11} when masses are measured in kilograms (Kg) distances in metres (m) and time in seconds (s)[5].

The calculation of V for the Sun is slightly more complicated because we have to add up the contributions of all gas particles in a gravitational field that they all generate together. Considering that the Sun is to very a good approximation a sphere, and assuming for simplicity a constant density throughout its interior we have that $V = -(3/5)GM^2/R$, where M is the mass of the Sun[6] and R its radius[7].

Before determining the age of the Sun under these assumptions, we may ask ourselves what happens to the star when its reservoir of gravitational potential energy is depleted because of the energy lost from the surface. If V decreases – meaning it becomes more negative – and given that the mass of the Sun stays constant, R must decrease, hence the Sun was expected to contract over time. Helmholtz calculated the amount of shrinkage per year necessary to replace the lost energy, finding that it was of just about 70–80 metres.

Using the expression for V given above and the present luminosity of the Sun assuming it does not change with time, the total solar lifetime turns out to be of about 20 million years. This is the result obtained by

---

[5]The notation 10^{-11} is a compact way to write 0.00000000001, with the number 1 occupying the 11th decimal place to the right of the decimal point. It is called exponential notation and is practical when dealing with very large and very small numbers. In general, 10^l represents the number 1 followed by l zeros, while 10^{-l} represents a number with the digit 1 occupying the l-th decimal place to the right of the decimal point.

[6]The mass of the Sun is equal to 1.988×10^{30} Kg, about 333,000 times the mass of the Earth.

[7]The solar radius is equal to about 695,700 Km, about 109 times the radius of the Earth.

Helmholtz, a value much smaller than Darwin's geological estimate of the age of the Earth. Similar conclusions were also reached by William Thompson, later Baron Kelvin of Largs[8], one of the greatest physicists of the nineteenth century.

Kelvin, like Helmholtz, was convinced that the solar energy reservoir was gravitational energy and in Macmillan's Magazine, vol. 5 (March 5, 1862), he gave an age of about 20 million years consistent with Helmholtz result[9], and also estimated the effect of a more realistic non-uniform density within the Sun (an increasing density towards the centre that increases the available gravitational energy for a fixed mass and radius, compared to the case of uniform density). He finally concluded that it was most probable the Sun was younger than 100 million years, and almost certainly it was younger than 500 million years.

Darwin was troubled by Kelvin estimate of the age of the Sun, which made very improbable the age of the Earth he estimated in his book. In a letter to the geologist James Croll dated 31 January 1869, Darwin expressed his worries about the short duration of the world according to Kelvin, in contrast with the very long time required by his theory, and eventually removed any reference to timescales from later editions of *Origin of the Species*.

In the end Kelvin was wrong regarding the age of the Sun and the Earth, that is today fixed to about 4.5 billion years (more on this in chapter 7), but of course he could draw conclusions only based on the knowledge at the time, when nuclear reactions (that power the Sun) were unknown. The point I wish to make with this example is that the estimate of the age of the Sun had a direct bearing on other, in principle completely different fields of research, like geology and biology.

1.3   CONFLICT OVER THE AGE OF THE UNIVERSE

An age discrepancy qualitatively similar to the Darwin-Kelvin disagreement over the age of the Earth was debated during the second half of the 1980s and the first half of the 1990s. The debate is nicely summed up by the title of a 1995 paper published in the journal *Nature* by Michael Bolte and Craig Hogan: *Conflict over the age of the universe* [15]. Even more dramatic was another *Nature* paper published the following month, titled: *Big Bang not yet dead but in decline* [73].

---

[8]The Kelvin is a small river that flows near Glasgow University.

[9]Incidentally, Kelvin thought that the Sun was an incandescent liquid.

We have two independent routes to the determination of the age of the universe. The first one rests on our current cosmological model, the so-called Big-Bang cosmology discussed in the next chapter. The second route is based on the theory of stellar evolution and the age of stars. Given that the universe cannot be younger than its stars, the age of the oldest stars sets a firm lower limit to the age of the cosmos.

Let's start with the age of the universe determined from cosmology. About 90 years ago it was discovered that the universe is expanding, with a velocity that increases with the distance (see the next chapter). More distant galaxies appear to move away from us with increasingly larger velocities v, with v being equal to the product $H_0 D$, where D is the distance and H_0 is a coefficient, a measure of the expansion rate of the universe, called Hubble constant. The value of H_0 is the same across the universe at a given time, but changes with time[10]. Its rate of change depends on how much matter and energy[11] are contained in the universe. If we know H_0 and the density of matter and energy in the universe now, we can determine how the expansion rate has evolved in the past to reach the present value. In this way we can determine how long ago the expansion has started, the moment of the Big-Bang, the 'birth' of the universe, so to speak. Let's denote with t_{bb} the age of the universe determined this way, that rests on the assumption that the Big-Bang theory is correct, and on the determination of the present values of the Hubble constant and the universe matter/energy density.

As for stellar ages, the age of the oldest stars has been traditionally based on the determination of the age of samples of globular clusters of the Milky Way, starting about 70 years ago with the pioneering efforts of Martin Schwarzschild, Fred Hoyle, Allen Sandage, and Halton Arp.

The Milky Way hosts about 150 globular clusters, ball-shaped collections of stars (typically around 1 million), tightly bound by their reciprocal gravitational attraction and distributed spherically around the core of our galaxy. Messier 13 (M 13 in brief) in the constellation of Hercules – shown in Fig. 1.2 – is one of the brightest globular clusters, and can be picked out with binoculars. The distribution of the luminosities and colours of their stars suggest very old ages, as I will

---

[10]Readers will have noticed that we are essentially assuming a common time for the whole universe. This sounds strange, given that after Einstein theory of relativity we are used to thinking that time cannot flow with the same speed across the whole universe. More on this in the next chapter.

[11]Einstein's special theory of relativity tells us that anything having a mass has also an equivalent amount of energy. The opposite is also true (see the appendix).

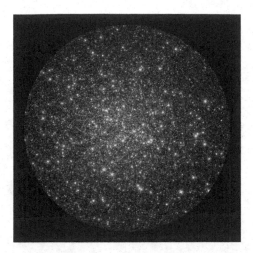

Figure 1.2 The globular cluster M 13 (NASA, ESA, and the Hubble Heritage Team STScI/AURA). Each star is bound to the cluster by the joint gravitational attraction of all other stars. Stars loop inward and outward with velocities large enough to balance the gravitational attraction.

show in chapter 7, and to date, ages of about 60 globular clusters have been determined by various researchers, including myself.

Figure 1.3 highlights the conflict over the age of the universe that was raging at the beginning of the 1990s. The filled circles display estimates of the ages of 24 globular clusters as a function of a measure of their initial chemical composition. The quantity [Fe/H] will be introduced in chapter 6, and it is a measure of the logarithmic abundance ratio between iron (Fe) and H relative to the Sun[12]. A value [Fe/H]=0 means an initial chemical composition like the Sun, while [Fe/H]=−2 denotes a composition with 100 times lower abundance of iron. This element is synthesized in the core of exploding stars that we name supernovae (see chapter 4), and it is injected in the interstellar medium during the explosion.

The little bars protruding from each circle span the full range of ages obtained for each cluster, when considering the errors on the age

---

[12]I will use base-ten logarithms throughout the book. I recall here that the logarithm of a number x ($\log(x)$) is the power to which the base (the number 10 in our case) must be raised to obtain x. For example, $\log(1000)=3$ because $10^3=1000$, and $\log(0.001)=-3$ because $10^{-3}=0.001$.

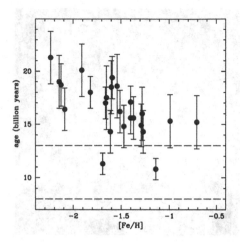

Figure 1.3 Ages of a sample of Milky Way globular clusters (filled circles with error bars) estimated in 1992 [18]. The two dashed lines bracket the range of possible values of t_{bb} from the accepted cosmological model at the time. The horizontal axis displays a measure of the initial chemical composition of the stars in the various globular clusters. Lower values of [Fe/H] mean lower amounts of iron (Fe).

measurements. The dashed lines bracket the range of possible values of t_{bb}, according to the uncertainty on the Hubble constant and matter/energy density at the time.

The existence of a major inconsistency between t_{bb} and globular cluster ages is undeniable and was very worrying. The oldest globular clusters appeared older than the universe, even taking into account the errors on their age. All researchers in the field, like Don Vanden-Berg and collaborators, Bruce Carney, Alessandro Chieffi and Oscar Straniero, Brian Chaboyer, Pierre Demarque and Ata Sarajedini – just to name a few – found similar results, with ages sometimes lower than the oldest values shown in Fig. 1.3, but the oldest clusters were never younger than about 16 billion years. For cosmology, theoretical arguments and measurements of H_0 strongly favoured the lower limit of the range of t_{bb} shown in the figure. I remember at the time I was once asked by a renowned cosmologist how much I could bring down the ages of globular clusters. I was young, with just one paper under my belt (on globular clusters, including also age determinations) and I did not have a coherent answer.

This severe discrepancy spurred people working on the mathematical models that describe the structure and evolution of stars ('stellar models') and age determinations, to check and refine both their calculations and methods. At the end of 1995, I moved to the Max Planck Institute for Astrophysics (MPA) in Garching near Munich, to work with Achim Weiss on globular cluster ages, trying to solve this discrepancy from the 'stellar' side.

The MPA is one of many research institutes of the German Max-Planck Society, all primarily devoted to fundamental research; its focus is the theoretical research in several fields of astrophysics, from stellar structure and evolution to the large-scale structure of the Universe and cosmology. It was established in Munich in 1958 under the direction of Ludwig Biermann, and under the second director, Rudolf Kippenhahn, the MPA moved to a new site in Garching, just north of the Munich city limits, in an area with a high concentration of scientific institutes. The Max Planck Institute for Extraterrestrial Physics is physically connected to MPA buildings, and about 50 metres away from MPA rise the headquarters of the European Southern Observatory (ESO). ESO is an intergovernmental European astronomy organisation founded in 1962 and currently supported by 16 European Countries plus Chile, that host the ESO facilities, amongst the largest and most (scientifically) productive telescopes in the world. It is currently building a 39-metre telescope which will become the world largest telescope (see chapter 10). Within walking distance from MPA you can encounter the Max Planck Institute for Quantum Optics and the Max Planck Institute for Plasma Physics. More recently, the scientific and engineering departments of the Technical University of Munich have also been relocated in the area.

At the time a path that first crossed large cornfields, and then followed the course of a narrow stream, led from the village of Garching to the research campus. It was an invigorating walk, and in the dark afternoons and evenings of winter, the snow and ice reflected the moonlight creating an almost magical atmosphere. It also wasn't uncommon to meet along the way people going or coming back from the institutes even at midnight or in the early morning hours.

The work at MPA was exciting, because of the computing facilities, the library, the vibrancy of the scientific discussions, the calibre of staff, postdocs and visitors, and the social activities. For me, one of the highlights was the Wednesday afternoon outdoor football game played on a full-sized pitch close to the cafeteria of the campus, that started invariably at 5 PM and lasted until physical exhaustion of the participants (a two-hour game at least... even astrophysicists can be physically very fit). We played with every weather condition (and with the illumination on in winter): In the dry summer heat, with rain and mud, with soft snow on the pitch, and even when the pitch was fully covered with ice[13], a health and safety nightmare. Achim and I struck a very good partnership as defenders (I usually played centre-back or right-back, and he typically played left-back).

After about one year of work on improving our theoretical models of stars, and devising the best possible theoretical approach to the age determination of globular clusters (more on this in chapter 7) we published in 1997 two papers (Scilla Degl'Innocenti, also at MPA at that time, worked with us on the first paper) with our new results [103, 104]. In the first paper, with the exciting title *The age of the oldest globular clusters*, we studied three 'classical' clusters (M 15, M 68, M 92), while in the second one we extended our analysis to a sample of 25 clusters. The objects of Fig. 1.3 were also amongst those analysed again by us. And this time we found that the ages were equal to at most 13 billion years!

I still remember vividly when I finally had models and tools fully developed and applied them to measure the age of the cluster M 68. It was around the end of a snowy winter when for the first time in my life I tried skiing; actually it was cross-country skiing, more akin to running, my sport of choice, compared to downhill skiing. When I got an age below 13 billion years I was thrilled, but also very worried that I made some major mistake. I checked and counterchecked several times, and the excitement grew exponentially; eventually, I talked to Achim and gave him the news. Our 'younger' globular cluster ages were confirmed shortly after by an independent route (more on this in chapter 7) based on results of the *Hipparcos* satellite, which measured the parallax (from which the distance is derived using simple geometry, see the appendix) for a huge sample of stars [41]. In the end, this 'conflict over the age of the universe' has been very beneficial, because it spurred us to revise and improve stellar evolution models, and the methods we used to determine globular cluster ages.

In a short article I wrote for the 1997 annual report of MPA, I concluded, full of youthful enthusiasm, that for the first time the age

---

[13] At first we played with a standard white football, hard to spot on a completely white pitch.

of the universe as determined from the age of its oldest stars agreed with the age determined from the Hubble constant.

On the other side of the 'conflict', cosmology also evolved. In 1998 Adam Riess, Saul Perlmutter, and Brian Schmidt discovered that the expansion of the universe is accelerating (see the next chapter)[14]; this was followed by the precise measurements of the tiny temperature fluctuations of the cosmic microwave background radiation (see the next chapter for more details) thanks to the Wilkinson Microwave Anisotropy Probe (WMAP) satellite launched in 2001 [52], later refined by measurements with the *Planck* satellite launched in 2009. These discoveries have improved the determination of the cosmological parameters, leading to a value of t_{bb} equal to about 13.8 billion years, roughly 1 billion years higher than the upper limit of the range shown in Fig. 1.3, and now consistent with the lower limit to the age of the universe set by the globular clusters.

GLOSSARY

Big-Bang cosmology: The currently accepted model for the evolution of the observable universe that describes how the universe has expanded from a very high-density and high-temperature initial state.

Cosmic microwave background radiation: Electromagnetic radiation that is the remnant from an early stage of the universe. It was released when the universe was about 380,000 years old.

Hubble constant: Parameter that describes how fast the universe is expanding at different distances from a particular point in space.

Globular cluster: Spherical collection of stars tightly bound by gravity that orbit the core of a galaxy.

Gravitational potential energy: Energy of an object due to its position in a gravitational field.

Radiotelescope: Specialized telescope made of an antenna and a receiver to detect radio waves from astronomical objects.

Stellar Parallax: Apparent shift of the position of a star against the

---

[14]They were awarded the Nobel Prize in 2011 for this discovery [94].

background of distant objects, due to the orbital motion of the Earth. Once a star's parallax is known, its distance from Earth can be easily computed from trigonometry.

FURTHER READING

Y. Lebreton. Stellar Structure and Evolution: Deductions from Hipparcos. *Annual Review of Astronomy and Astrophysics*, 38:35, 2000.

M. Salaris, M. and A. Weiss. Homogeneous age dating of 55 Galactic Globular Clusters, Clues to the Galaxy Formation Mechanisms. *Astronomy & Astrophysics*, 388: 492, 2002.

A. Stinner and J. Teichmann. Lord Kelvin and the Age-of-the-Earth Debate: A Dramatization, *Science & Education*, 12:213, 2003.

D.A. Vandenberg, M. Bolte and P.B. Stetson. The Age of the Galactic Globular Cluster System. *Annual Review of Astronomy and Astrophysics*, 34:461, 1996.

CHAPTER 2

An Echo of the Distant Past

Before embarking on our journey to discover how stars work and how we can determine their ages, it is important to understand the environment in which they form and evolve. This chapter summarizes our current view of the universe, how we describe its structure and evolution in time, and the role stars play in the making of the cosmic tapestry. The reader is referred to the appendix for more details about some of the physics mentioned in this chapter.

The vastness of space and lengths of time involved in the description that follows are literally mind-numbing and, to be honest, virtually impossible to relate to our everyday experience. Distances of millions or billions of light-years, timescales of billions of years, are way beyond our experience; celestial bodies are impossibly distant from each other, even within our own galaxy, and the pace of events in the universe is incredibly slow, bar sudden catastrophes like the death throes of massive stars, or the merging of compact objects like white dwarfs, neutron stars, and black holes (see chapter 4). Sometimes I feel that I can get a very partial awareness of this immensity when listening to the spectral and otherwordly mellotron of *Mysterious semblance at the strand of nightmares* from the Tangerine Dream's album *Phaedra*, or when the final *Celestial voices* section of Pink Floyd *A saucerful of secrets* starts, with the grandiose organ slowly cycling through that sublime chord sequence...

2.1 GALAXIES

Stars are not scattered at random across the cosmos, without discernible patterns, instead they are assembled into large structures that we call galaxies, like our own Milky Way. Today the existence of galaxies seems obvious, but it is by no means trivial. The evidence that stars are assembled into galaxies – and within a galaxy, a fraction of its stars are grouped into much smaller (in both mass and size) units, the star clusters – is a consequence of the forces at play, and how the universe has evolved. It is also extraordinary to think that it is only about a century ago when we have finally discovered that the Milky Way is not the entire universe, just one of a multitude of galaxies spread across the cosmos.

To try and get, at least partially, a sense of the immense size of the universe, let's consider that typical distances between stars in a galaxy are on the order of 1 parsec. Parsec (pc) is the unit of distance usually employed is astrophysics, and is equal to 3.26 light-years, where one light-year is the distance travelled by light – in vacuum – in one year, corresponding to about 9.4607×10^{15} metres (the speed of light is equal to 2.998×10^8 metres per second). The symbol Kpc is employed to denote a Kiloparsec, one thousand parsecs, and Mpc stands for a Megaparsec, corresponding to one million parsecs.

The stars closest to the Sun are Proxima Centauri, Alpha Centauri A, and Alpha Centauri B, at distances of about 1.3 pc. These distances are already truly enormous and hard to envision, if we think that in terms of light-travel time the Sun (at about 150 million kilometres from Earth) is at a distance of a mere 8 light-minutes. The most distant spacecrafts sent out from Earth are the Voyager 1 and Voyager 2, both launched in 1977, and currently travelling at a distance of 147 and 122 times the Earth-Sun distance, respectively, equal to just 20 and 17 light-hours.

Distances from one galaxy to another are much larger, typically between 100 Kpc and 1 Mpc. The Large and the Small Magellanic Cloud (LMC and SMC), which are practically in our backyard and visible to the naked eye in the Southern sky, are at distances of about 55 and 80 Kpc, respectively. The Andromeda galaxy, just about visible during clear nights in the Northern sky, sits at about 800 Kpc from us.

Observations tell us that there are three main types of galaxies: Spirals, ellipticals, and irregulars (see Fig. 2.1). The spiral galaxies – like the Milky Way – show generally a bright nucleus, the 'bulge',

Figure 2.1 *Left panel:* The spiral galaxy Messier 51 (M 51, 'The Whirlpool Galaxy'), about 10 Mpc away from the Sun (NASA/Hubble). *Middle panel:* The elliptical galaxy Messier 49 (M 49), located at a distance of about 17 Mpc from the Sun (ESA/Hubble & NASA, J. Blakeslee, P.Cote et al.) *Right panel:* The irregular galaxy NGC 5264, about 5 Mpc away in the constellation of Hydra (ESA/Hubble & NASA).

surrounded by a flat disk that contains gas, dust, and the luminous spiral arms. Some spirals have arms that are wound tightly, while others have very loosely wound arms. An extended faint, spherical halo, surrounds disk and bulge. The star clusters populating the halo of the Milky Way are the globular clusters introduced in the previous chapter. Star clusters orbiting the disk of the Milky Way are called open clusters, are usually younger and host fewer stars than globulars. The typical total mass of all the stars in a spiral galaxy is about $10^{11}\,M_\odot$ (a truly huge number) where the symbol $M_\odot$ denotes the mass of the Sun.

Apart from how much the spiral arms are wound around the nucleus, two broad categories of spiral galaxies do exist, normal and barred spirals. The classification depends on whether a bar springing symmetrically from the nucleus is present. In barred spirals (like our galaxy) the arms start at the ends of the bar whilst in normal spirals they originate from the nucleus.

Elliptical galaxies display shapes ranging from almost circular to long and cigar-like, show no disk, no spiral arms, and in general they resemble the bulges of spiral galaxies. Ellipticals have a large spectrum of sizes and masses, the total mass in stars ranging between about 10^7 and $10^{12}\,M_\odot$.

The link – in terms of morphology – between spirals and ellipticals is provided by the so-called lenticular galaxies that appear like elongated ellipticals with a bulge, but without bars and spiral arms. The third main class of galaxies are the irregular galaxies (the Magellanic Clouds are examples of irregular galaxies) rare and relatively faint, that show no regular structure and contain large amounts of gas and dust.

Even galaxies – like stars – are assembled into larger units, called groups and clusters. Groups typically host up to about 50 galaxies, have diameters of a few Mpc, and contain at least 50% of the galaxies in the local universe. Clusters of galaxies host from a few hundred to several thousand galaxies and have diameters generally between 2 and 10 Mpc.

The Milky Way – usually referred to also as the Galaxy – belongs to a galaxy group named, perhaps not surprisingly, Local Group. The Local Group is made of over 50 objects (mainly galaxies small in both size and total mass of stars, named dwarf galaxies) among them the LMC, SMC and Andromeda. The Milky Way and Andromeda are the two largest members of the Local Group, both spirals, each of them with a cohort of satellite galaxies. As an example, the Magellanic Clouds are satellites of the Milky Way, and the dwarf elliptical galaxy Messier 32 (M 32) is a satellite of Andromeda. These two galaxies with their cohorts move towards each other at a speed of about 120 Km/s. The total diameter of the Local Group is of about 3 Mpc, beyond which a few other galaxy groups bridge the distance to the Virgo cluster of galaxies about 20 Mpc from us. Beyond the Virgo cluster, dozens of more galaxy groups are scattered across the cosmos, until we reach the Coma cluster, at a distance of about 100 Mpc, hosting thousands of galaxies.

Deep (in the sense of reaching very faint and very distant objects) galaxy surveys have studied and are still probing the distribution of galaxies in the universe, and have revealed even more complex structures. Clusters of galaxies are grouped together in superclusters containing dozens of clusters, and extending for tens of Mpc. Superclusters, in turn, are arranged in filaments and sheets separated by large voids. On these large scales, the distribution of the matter we can see resembles a foam-like structure, the so-called cosmic web.

In addition to the matter we can 'see' (stars, gas, dust) through the emission of electromagnetic radiation (light in various wavelength ranges), the universe contains an enormous amount of so called dark matter, that does not release electromagnetic radiation. First indications for the presence of dark matter emerged during the 1930s, after Fritz Zwicky measurements of the velocities of galaxies belonging to

Figure 2.2 The galaxy cluster Abell 2537 in the constellation of Pisces (ESA/Hubble/NASA).

the Coma cluster, taking advantage of the Doppler effect, something routinely experienced in our everyday lives.

When we listen to the sound of a moving object, like the siren of a police car, we can notice that if the car is approaching us the pitch of the sound gets increasingly higher (wavelengths getting shorter) because the soundwaves we hear are being squashed (their wavelengths compressed) in the front of the vehicle, that is literally catching up with its own sound. The pitch gets lower (wavelengths gets longer) if the car drives away from us, for the soundwaves now are more spread out because the siren is moving away from the sound that is reaching us. The same is true also for the light given off by an object approaching (shifted to shorter wavelengths, or 'blueshifted') or receding (shifted to longer wavelengths, or 'redshifted') along our line of sight. The amount of blueshift or redshift depends on the speed relative to us. In the case of stars and galaxies this effect can be measured by comparing the wavelengths of the lines in their spectra, with those of a source at rest (see the next chapter and chapter 9).

The velocities measured by Zwicky are determined by the strength of the gravitational attraction felt by the galaxies, which in turn depends on the total mass contained in the Coma cluster. In his analysis he found a cluster mass larger than $4.5 \times 10^{10} M_\odot$, and then used the typical luminosity of a galaxy as estimated at the time – about $8.5 \times 10^7 L_\odot$ (where $L_\odot$ is the luminosity of the Sun) – to calculate the ratio γ between the total cluster mass and the sum of the luminosities of all its galaxies. This ratio[1] turned out to be around 500 or more. On the other hand, estimates of γ considering stars within about 8 Kpc from the Sun in the disk of the Milky Way – they were called Kapteyn system at the time – gave $\gamma=3$. By assuming that the stars in the Kapteyn system had masses and luminosities similar to those of the stars in the Coma galaxies, Zwicky's result suggested the presence of much more mass within the cluster than what is locked in their stars [128].

Later on, during the early 1970s, Vera Rubin and Kent Ford measured the velocity of cold hydrogen gas clouds, that orbit around the disk of spiral galaxies (like stars do) at a range of distances from the centre [97]. These clouds emit electromagnetic radiation with a wavelength of 21.1 centimetres – a 'long' wavelength detectable by radio-telescopes – caused by the change of the spin of the electrons around the hydrogen nuclei. In a hydrogen atom the single electron that 'orbits' around the proton can have a spin (the spin is a quantum property that we can picture as rotation of a particle about its axis) in the same direction as the spin of the proton (denoted as parallel) or in the opposite direction (anti-parallel). The electron spinning anti-parallel has slightly lower energy than the case of parallel spin hence, given that nature aim is always to reach the state of lowest possible energies, the parallel spin electrons will eventually flip to the anti-parallel direction, the energy difference being released as electromagnetic waves with a wavelength of 21.1 centimetres. The analysis of the Doppler shifts of this line provides a measure of the orbital velocities of the gas clouds.

It is expected that the clouds outside the visible edge of a spiral galaxy would be moving slower than gas at the edge if the mass is concentrated where the galaxy emits light, for the gravitational pull of the galaxy would be weaker on these more distant clouds. Instead, Rubin and Ford found that the orbital velocity remained constant outside the visible edge, suggesting the presence of additional dark (invisible) matter, whose amount increases with increasing distance from the centre.

---

[1]Zwicky's calculation of the mass of the cluster was significantly underestimated, but also the typical luminosity of a galaxy was underestimated.

The signature of dark matter is also found within clusters of galaxies by studying the intracluster medium, namely the hot gas distributed amongst the galaxies (at temperatures on the order of 10–100 million Kelvin[2]), made of mainly hydrogen and helium nuclei that have lost their electrons (we say they have been ionised) due to the high temperatures. These nuclei are electrically charged particles (because of the positive charge of their protons) accelerated by the gravitational pull of the cluster, and they emit electromagnetic radiation, like any electric charge that is accelerated in a science lab here on Earth. The wavelength of this radiation (typically X-rays, at very short wavelengths) depends on the acceleration of the gas which, in turn, depends on the strength of the gravitational attraction, hence the amount of mass within the cluster. The mass distributions determined from these X-ray observations give values that far exceed the luminous mass, and indeed reveal the presence of large amounts of dark matter.

Yet another sign of the presence of dark matter in the universe is the appearance of rings and arcs in images of galaxies. They can be explained if the light from distant galaxies is being distorted and magnified, like going through a lens, by massive concentrations of dark matter in the foreground. This phenomenon, the 'gravitational lensing', is a consequence of the fact that mass bends the space: This is the origin of the force of gravity, according to Albert Einstein general theory of relativity. The path of light rays coming from luminous objects follows the curvature of space, and if the objects are located beyond large concentrations of matter, the image is bent and distorted towards the observer, just like an ordinary lens.

Due to this gravitational lensing, we often record images of distant galaxies that are stretched or magnified, and even multiple images of the same galaxy. The analysis of the image patterns of the lensing tells us about how dark matter is distributed in space.

I hope it has transpired from this very brief overview that matter in the universe appears to be clumpy, organized hierarchically into structures of increasing sizes, with vast regions of empty space in between. However, if we average the properties over volumes of space with radius on the order of, let's say 100 Mpc (a very small distance

---

[2]Kelvin is the unit of temperature measurement used physics. Temperatures on this scale can be transformed to the more familiar Celsius degrees (C) by subtracting the number 273 to the temperature given in Kelvin.

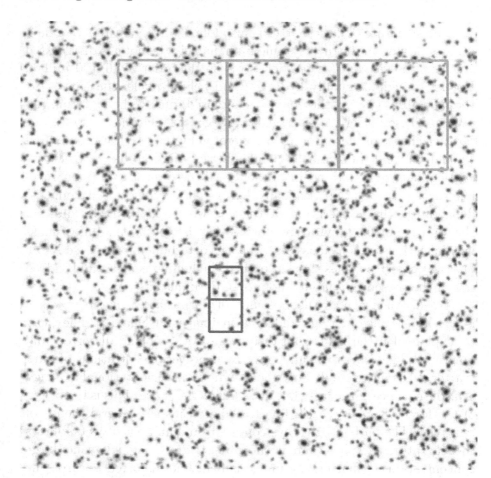

Figure 2.3 Visualization of the cosmological principle. Squares of different dimensions are superimposed to a background of galaxies. The average number of objects in each square varies among the smaller ones, whilst it is pretty much the same when bigger squares are considered (courtesy of David Hyder).

compared to the size of the observable universe), the universe appears much smoother, in the sense that the content of each spherical volume of this radius is on average the same – provided we don't reach too large distances, because we would be looking too far back in the past due to the finite speed of light (see below), and potential evolution of the universe with time might have changed its appearance. A qualitative example is displayed in Fig. 2.3, which shows a background of galaxies, with highlighted square-shaped areas of different dimensions. The distribution of galaxies is uniform across the whole picture but, if we consider small areas, the number of objects can change a lot from one location to the other. In the case of the two small squares, the number of galaxies they enclose is very different. However, when we take bigger regions more representative of the large scale distribution of galaxies, but still small compared to the size of the whole picture, the result changes. If we consider the three larger squares, the distribution of galaxies is on average the same and it is a much better reflection of the distribution across the whole figure.

This observed smoothness ad uniformity of the distribution of galaxies is enshrined in the cosmological principle. Stated formally, the cosmological principle says that on large enough scales – but still small compared to the size of the universe – the distribution of matter in space is homogeneous (the same at different locations) and isotropic (the same when looking in any directions) at any time.

Another important piece of information gathered from observations of populations of galaxies at increasing large distances, is that they evolve in time. Not just the luminosity, colour, and size of a galaxy, but also the fraction of galaxies of different types vary with time. Observations of distant galaxies show us how they appeared when their light was released. This phenomenon is called lookback time, and affects any observation; For example, when we observe the Sun, we see it as it was about eight minutes earlier.

There is an important issue to clarify at this point. Back in time means far out in space, hence observing at different lookback times means to sample different regions of the universe. But according to the cosmological principle[3] all regions of the universe are equivalent at any time, and what we see at different lookback times mirrors the evolution of any single large patch of the universe, including the one we inhabit.

---

[3]We will see shortly how our model for the global evolution of the universe, based on the cosmological principle, explains all the relevant observations.

2.2 EXPANSION OF THE UNIVERSE AND COSMIC MICROWAVE BACKGROUND RADIATION

Between 1912 and 1922 Vesto Slipher measured the spectra of a large sample of spiral galaxies[4], and found that for most of them the spectral lines were redshifted. Let's quantify this redshift by the ratio $z = (\lambda_{obs} - \lambda_s)/\lambda_s$, where λ_{obs} is the observed wavelength of a given spectral line, and λ_s its value at rest. A value z=0.01 for example, means that the observed wavelengths have been redshifted by 1%.

If Slipher measurements of z are interpreted as Doppler effect, the implication is that these galaxies are moving away from us. But this conclusion was at first too extreme for a time when it was customary to believe that the universe was static, and it wasn't even sure that these objects were stellar systems beyond the Milky Way[5].

Knut Lundmark found in 1924 a general correlation between redshift measurements and distances of the spiral galaxies. The farther away the galaxies, the larger the redshift, but the scatter around this mean trend was huge, because of the large uncertainties of the distances. Three years later, the priest and cosmologist Georges Lemaitre obtained a clearer correlation using a different set of distances, but his paper published in a Belgian journal went unnoticed [71]. Finally in 1929, using yet improved – for the time – distance estimates, Edwin Hubble announced the distance-redshift relationship for galaxies which commonly carries his name [57], although recently the International Astronomical Union has recommended to name it 'Hubble-Lemaitre' law. For the galaxies included in Hubble and Lemaitre studies, with redshifts z smaller than 0.01, the Hubble-Lemaitre law can be written as $zc = H_0D$ where c is the speed of light, D the distance, z the redshift. The quantity H_0 is a number named 'Hubble constant': Due to systematic errors in his distances (they were all too short), the value of H_0 derived by Hubble was about seven times larger than what is accepted today. Figure 2.4 displays a more modern plot of the Hubble-Lemaitre law, for redshifts up to $z \sim 0.06$. It shows clearly the average linear relationship (the data are distributed along a straight line)

---

[4]At the time they were called spiral nebulae, because it wasn't yet clear that they were galaxies like ours.

[5]To give a flavour of the debate going on at the time, I mention here that in the conclusions of a 1919 paper titled *On the existence of external galaxies*, Harlow Shapley – the discoverer that the Sun is not at the centre of our galaxy– wrote that the evidence supporting the existence of other galaxies beyond the Milky Way appeared unconvincing.

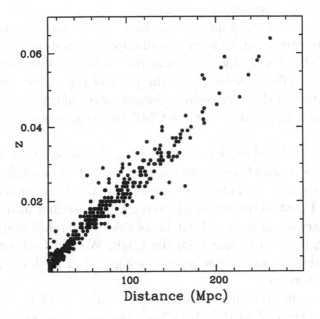

Figure 2.4 Hubble-Lemaitre law for a sample of galaxies out to distances of about 300 Mpc [72]. The vertical axis displays the redshift z, the horizontal axis the distance D of each galaxy. The points, that represent observations of individual galaxies, are distributed along a straight line.

between the measured redshift z and the distance D of galaxies, meaning that z increases with increasing D, but the value of the ratio between z and D is constant on average.

Works in the last two decades have extended measurements of this relationship out to distances about 1,000 times longer than the distances sampled by the studies of Lemaitre and Hubble, showing that this relationship is no longer a straight line over this much larger range [94].

Another discovery of fundamental importance for our understanding of the universe was made serendipitously in 1965 by Arno Penzias and Robert Wilson, while working with a 20-foot horn-shaped antenna built by Bell Labs in New Jersey. They couldn't get rid of a persistent noise coming from all directions in the sky, even after having cleaned the antenna and removed pigeons (and their droppings) nested inside the horn, and had to conclude that it was a real signal. The strength of this signal – called cosmic microwave background (CMB) radiation – as a function of the wavelength λ shows a maximum at λ equal to about 2 millimetres, and a steep decrease at longer and shorter wavelengths. This behaviour turns out to be very well approximated by the radiation given off by a blackbody, like the radiation inside an oven (see chapter 3 and chapter 6). The wavelength of the peak of the signal depends on the temperature of this virtual oven, longer wavelengths corresponding to decreasing temperatures; for the CMB this temperature is equal to just 2.725 K[6].

When we still had analogue televisions, the 'snow' that appeared on screen in between channels was made up of interference from background signals that the antennas were picking up. I remember that as a kid I found that 'snow' quite frightening, like peeking into a forbidden, scary parallel universe. About 1% of this interference (and I didn't know this at the time) came from the CMB. We will find out shortly what I was actually looking at when staring at that frightening screen on the television set.

After removing the effect of the motions in space of the Sun (that orbits around the disk of the Milky Way) and of our galaxy, the CMB temperature is to a very good approximation uniform across the whole sky, a fact very hard to explain if the CMB is produced by discrete sources, like stars or galaxies [87].

More recently, the COsmic Background Explorer (COBE) satellite discovered in 1992 tiny variations of the CMB temperature when looking at different points in the sky[7]. These detections have been refined later on by the Wilkinson Microwave Anisotropy Probe (WMAP) and the Planck satellite measurements. Figure 2.5 shows a depiction of the blackbody temperature of the CMB over the full sky (projected onto an oval, similar to a map of the Earth): Fluctuations are extremely small, on the order of typically 0.001%, but present. By computing the average over the sky of the temperature fluctuation measured from any two points separated by an angle ϕ, we obtain what is called the angular power spectrum of the CMB temperature fluctuations, also shown in Figure 2.5. This power spectrum shows the existence of a series of peaks and troughs located at specific values of ϕ.

---

[6]In 1978, Penzias and Wilson were awarded the Nobel Prize for this discovery.

[7] John Mather and George Smoot were in charge of the instruments on board COBE, and were awarded the Nobel Prize in 2006 *for their discovery of the black-body form and anisotropy of the cosmic microwave background radiation.*

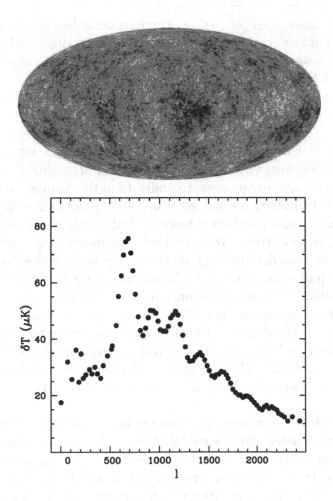

Figure 2.5 *Upper panel:* Variation of the temperature of the CMB across the sky (shown as grey-scale. Darker grey means higher temperatures) from the WMAP satellite measurements. The range of the fluctuations δT around the mean temperature of the CMB (equal to 2.725 K) is on the order of 0.001% (NASA/WMAP Science Team). *Lower panel:* Power spectrum of the temperature fluctuations [90]. The vertical axis displays δT in μK (1 μK corresponds to 10^{-6} K), while the horizontal axis displays the wave number l=180/ϕ, where the angle ϕ is measured in degrees.

2.3   THE BIG-BANG COSMOLOGICAL MODEL

How do we make sense of the observations described in the previous sections? If at this stage you think that even just trying to address this question is insane, you are in good company. The great ancient Greek philosopher Socrates was reported (by his pupil Xenophon) to have said that speculators on the universe and on the laws of the heavenly bodies were no better than madmen.

The key to address this question without – maybe – being labelled as 'madmen', is to make the assumption that a unique set of laws of physics applies everywhere in the universe. Without this, it would be impossible to make any progress to understand the cosmos. To date, all analyses of the radiation collected from the most distant objects have not shown any indisputable evidence that the laws of physics change in different parts of the universe, or that they have changed with time.

When we consider the huge distances among galaxies, and given there is no indication that matter in galaxies has a net electric charge (either positive or negative) the only fundamental force of nature able to shape the structure of the universe is gravity, described by Einstein general theory of relativity. Electromagnetism, the other long-range fundamental force[8], cannot play any role on large scales if galaxies are electrically neutral, whilst the weak and strong forces (the last two fundamental interactions) act only over subatomic distances (see the appendix).

In 1917, barely two years after the publication of his general theory of relativity, Einstein built a model of the universe assuming a static and uniform distribution of matter[9]. He had no indications that the universe might be expanding or contracting, and to keep it static he needed to add to his gravitational field equations a positive constant – called cosmological constant – and tune appropriately its value. The ratio between the density of matter and the cosmological constant had to have a very specific value: Any infinitesimal change of this ratio would result in a universe that either expands or collapses, as demonstrated first by Willem de Sitter and later by Arthur Eddington. The same year de Sitter himself built a model of the universe without matter, but with a positive value of the cosmological constant, believing it

---

[8]'Long range' means that the force can act over indefinitely large distances.

[9]A uniform distribution of matter was assumed to simplify the solution of the equations, but also to avoid a universe in which all viewpoints are not equivalent. This assumption is essentially the same as the cosmological principle.

was static like Einstein universe. It was realized later, by recasting in a different form the mathematical description of the geometry of space, that this model universe actually expands.

A more systematic analysis of universes made of a homogeneous and isotropic fluid (according to the cosmological principle) whose particles are galaxies started with the seminal works by Alexander Friedmann in 1922–24, Lemaitre in 1927, Howard Robertson and George Walker in 1935. These so-called called Friedmann-Lemaitre-Robertson-Walker models serve as the standard framework for all modern studies of cosmology, and can accommodate a universe (with matter and eventually also a cosmological constant) that expands from a very dense initial state, providing an explanation for the Hubble-Lemaitre law[10]. But how exactly does the expansion of the universe predicted by these models account for the Hubble-Lemaitre law?

The simplest explanation is that the galaxies themselves recede from us, hence the redshift is due to the well-known Doppler effect. For speeds much lower than the speed of light c, the redshift would then be equal to $z = v/c$, where v is the galaxy recession velocity, and the Hubble-Lemaitre law becomes $v = H_0 D$. When v approaches the speed of light we need to apply corrections according to the special theory of relativity, because galaxies cannot move faster than light: As a consequence, v can never reach c.

According to the Friedmann-Lemaitre-Robertson-Walker models the explanation for the Hubble-Lemaitre law, as surprising as it seems, is however that galaxies are not moving away from each other, instead the space between galaxies is literally stretching. Each galaxy can be considered to first approximation as a static object located at a specific point in space, while space itself – the distance between any two points – is expanding with a rate that is the same everywhere at a given time, as shown in Fig. 2.6. It is important to stress that there is no centre to the expansion, there are no edges, otherwise the cosmological principle would be violated. From any position in the universe, all points between an observer and a distant galaxy increase their separation at a constant rate, hence more distant galaxies recede from the observer with larger velocities.

The redshift is caused by the expansion of space that stretches out the travelling light waves during their journey to us, increasing their

---

[10]Actually Hubble in 1929, and as late as 1936, didn't interpret his law as a signature of the expansion of the universe.

wavelength. The more distant the galaxy, the faster the expansion of the universe at source, the longer the travel and the more stretched (more redshifted) the light waves we observe. This explains the observed Hubble-Lemaitre law. From this interpretation of the redshift we derive $v = H_0 D$, but this time the galaxy recession velocity v is actually the expansion velocity of space at a distance D from us. This same relationship between v and D also holds when v approaches and eventually exceeds the speed of light, because the galaxies themselves are not moving, it is the expansion of space that causes their recession.

The currently accepted value for H_0 is of about 67 Km per Megaparsec per second (Km/(Mpc s))[11], meaning for example that a galaxy at a distance of 0.5 Mpc is receding from us with a speed of 33.5 Km/s, and a galaxy at 10 Mpc is moving away with a speed of 670 Km/s.

As I hinted at in the previous section, not all galaxies conform to the Hubble-Lemaitre law, and the reason is simple. If objects are bound by their reciprocal gravitational attraction, like the planets in the solar system are bound to the Sun, they no longer recede from each other. On short distances, local gravitational attractions are stronger than the effect of the expansion of space; this is why distances within the solar system do not increase, or why the Andromeda and Milky Way galaxies are actually approaching each other, not expanding away. These so-called peculiar motions cause additional wavelength shifts in the spectra of galaxies according to the Doppler effect, because in this case the galaxies themselves are moving with respect to us. When we observe objects at very large distances, velocities due to the stretching of space get increasingly higher, and motions due to the effect of the local gravity eventually become negligible compared to the cosmic recession velocity.

The expansion of space with a rate that is the same everywhere is the consequence of the cosmological principle embedded in the equations of general relativity. As a bonus, we can use this property to define a common time for every observer in the universe. We know from relativity that observers in motion with respect to each other won't agree on whether two events are simultaneous (happen at the same time according to their clocks) or not. However, in a homogeneous and isotropic universe, we can define a so-called cosmic time as

---

[11]The determination of H_0 has a long history. Hubble first determination was of about 500 Km/(Mpc s) and for several years estimates were clustered around 50 and 100 Km/(Mpc s). Today there is some debate between values around 67 and slightly more than 70 Km/(Mpc s).

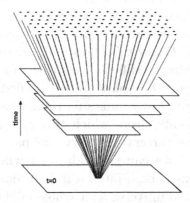

Figure 2.6 Qualitative two-dimensional depiction of the expansion of the universe. Galaxies are located at fixed points in space, and their distance increases with increasing cosmic time t, due to the expansion of space. At time zero all galaxies in the figure were contained in an infinitesimally small area (courtesy of David Hyder).

the time measured by a clock co-moving with the expansion of the universe, essentially a clock anchored to a given point in space. Two clocks co-moving with the expansion of the universe at different locations can be synchronized to a common time, when for example the local average density of matter around each clock has the same prescribed value. After this synchronization, when the clocks show the same time the average density of matter around each clock will always be the same (but possibly different from the value at synchronization). In this way, we can assign a unique timeline to the evolution of the universe, the same everywhere. This is why I could write words like 'at a given time' or 'at this time' in the previous discussions: I was referring (and will be referring in what follows) to the cosmic time.

Now, if we run the 'movie' of the expansion of the universe backwards, any region of the universe we can observe – no matter how large – must have expanded from an infinitesimally small volume of space. We can set the zero of the cosmic time to this moment that marks the start of the expansion of the universe, the 'Big-Bang', as named by the British astrophysicist Fred Hoyle during a talk for a BBC radio broadcast in 1949. Big-Bang is also the name commonly given to the cosmological model discussed in this section. It is, however, important to recall that despite the models point to a zero cosmic time of infinite density and virtually zero volume, our current physics theories cannot describe events happening earlier than 10^{-43} seconds after this hypothetical origin of time. Below this threshold, the distances are so short that a – currently lacking – unified description of gravity together with the three other fundamental interactions is needed (see the appendix).

Even if the term Big-Bang makes us think of a sort of explosion in space, there was no centre from which the matter started to expand outward. Also, it is the currently 'observable' part of the universe (the region of the universe we can see today – I will come back to this point shortly) that was packed into an infinitesimally small volume at the Big-Bang, while the universe as a whole might have been already infinite, extending beyond 'our' observable portion (see Fig. 2.6). To summarize, we can only say that at the start of expansion the whole universe was extremely dense, but it wasn't necessarily extremely small. Another important point to make is that the Big-Bang model does not explain how the universe came into existence in the first place. As mentioned before, our current description of physics does not apply to the time of the start of the expansion. We just assume that space, time, and energy already existed.

Having established that the universe is currently expanding from an initial state of virtually infinite density, it makes sense to ask whether the expansion will last forever, and also, more in general, what determines the past-, present-, and future speed of the expansion of space. The equations of the Friedmann-Lemaitre-Robertson-Walker models can answer this question.

The time evolution of the expansion rate of space is fixed by the type of matter and energy in the universe, the present values of their densities, and the value of H_0. Type of matter here means essentially whether the particles have zero or almost zero mass like photons – the particles representative of electromagnetic waves[12] – and neutrinos, or are massive, like for example electrons or protons. In the first case particles are 'fast', or 'hot', because they travel at the speed of light (the mass-less photons) or close to the speed of light; in the second case they move at much slower speeds and are denoted as 'cold'. The

---

[12]Electromagnetic waves of different wavelengths can be treated also as massless particles (photons) of different energies (energy decreases when the wavelength increases) according to the laws of quantum mechanics. The same laws assign to each particle – like protons or electrons for example – a wavelength named De Broglie wavelength.

energy is classified instead according to how its density changes with the expansion of the universe. We name it cosmological constant if the density stays constant, otherwise, if the energy density varies with the expansion of the universe, the term 'dark energy' is used. In a way, the cosmological constant can be seen as a special type of dark energy.

The measured density ρ_p of any type of particle is used to calculate the ratio $\Omega_p = (8\pi G\rho_p)/(3H_0^2)$, where G is the gravitational constant, and H_0 the Hubble constant, while the dark energy/cosmological constant contribution Λ enters the ratio $\Omega_\Lambda = (\Lambda c^2)/(3H_0^2)$. If the sum $\Omega = \Omega_p + \Omega_\Lambda$ is equal to 1 (where Ω_p includes the contribution of all types of matter, including photons and dark matter) the geometry of space is flat. This means that if we could draw instantaneously a triangle between three galaxies with side lengths on the order of 100 Mpc (the distances over which the universe can be considered homogeneous and isotropic) the sum of its three angles is equal to 180 degrees, the same as for triangles drawn on a flat surface like a sheet of paper. If Ω is larger than 1, the geometry of space is spherical, the three-dimensional counterpart of the two-dimensional surface of a sphere. The sum of the angles of our hypothetical triangle is more than 180 degrees, like for a triangle drawn on a globe, with one side along the equator and the other two sides following two longitudinal straight lines up to the north pole. Finally, if Ω is smaller than 1, the geometry is hyperbolic, as the surface of a saddle, and the same sum will be smaller than 180 degrees. Estimates of the present densities of the various components of the cosmic fluid and the Hubble constant show us that $\Omega=1$, Ω_Λ is different from zero and positive, and the density of fast particles (photons and neutrinos) is negligible:

- $H_0 = 67$ Km/(Mpc s);

- $\Omega_\gamma = 0.0001$ – photons;

- $\Omega_\nu = 0.001$ – neutrinos;

- $\Omega_M = 0.049$ – ordinary matter made of protons, neutrons, electrons, including planets, stars, galaxies, gas, dust, the 'visible matter';

- $\Omega_{DM}=0.27$ – dark matter;

- $\Omega_\Lambda=0.68$ – dark energy.

The dark matter particles whose nature is still unknown, are expected to be 'cold', otherwise numerical simulations for the formation of galaxies (more on this below) cannot reproduce their observed distribution in space. As for the dark energy, all indications point to a cosmological constant, whose density does not change with the expansion of the universe. The origin of this energy contribution is also still unknown. It is striking and sobering to realize that about 95% of the matter-energy content of the universe is of unknown origin.

The values $\Omega=1$ and $H_0 = 67$ Km/(Mpc s) correspond to a total density of matter plus energy[13] of about 10^{-26} kilograms (just about 6 hydrogen atoms) per cubic metre of space. To appreciate how empty the universe is, let's recall that the density of air at sea level is typically between 1.23 and 1.28 kilograms per cubic metre, about 26 orders of magnitude (a factor 10^{26}) larger than the average matter/energy cosmic density.

Let's now imagine to take a snapshot of a large portion of the universe at this cosmic time: Galaxies will be distributed in space at certain distances from each other. If we neglect the peculiar motions due to the local gravity, these distances have changed in the past and will change in the future because of the expansion of the universe. Figure 2.7 displays how they vary (by how much they are increased or reduced) as a function of time compared to the present values, for $\Omega=1$ and three different values of Λ, assuming the current estimate of H_0. The age of the universe since the Big-Bang is marked by the points along each curve with the vertical coordinate equal to 1.

The figure shows clearly how the past, present, and future evolution of the expansion rate of the universe (and its present age, the time elapsed since the Big-Bang) depend strongly on the value of Λ, for fixed Ω and H_0. In general, irrespective of the exact value of Λ, the expansion rate of the universe – the slope of the lines in this diagram – changes with time, and as a consequence the Hubble constant also changes with time. Its value is constant across the universe at a given cosmic time t, but changes with t. The symbol H_0 always denotes the present value.

The model with negative value of Λ ($\Lambda < 0$) predict an age of the universe of about 7.5 billion years, and that at some point the expansion of space will stop, followed by a contraction. During the contraction the

---

[13] An energy E can be transformed into an equivalent amount of matter m and vice versa, according to Einstein $E = m\,c^2$ relation, see the appendix.

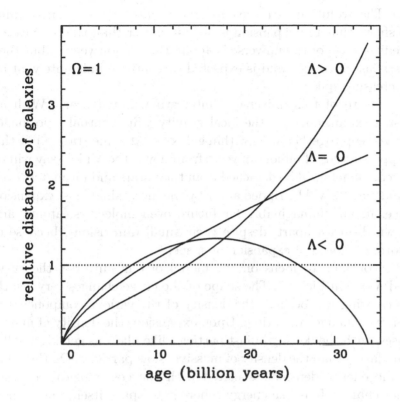

Figure 2.7 Time evolution of galaxy distances compared to the present values, for models with flat geometry, $H_0=67$ Km/(Mpc s), and three values of the cosmological constant Λ. The model shown for $\Lambda > 0$ corresponds to the currently accepted model of our universe ($\Omega_\Lambda=0.68$). Age (the value of the cosmic time) is counted starting from the Big-Bang when the galaxy distances were equal to zero. The current age of the universe predicted by these models corresponds to the points along each curve with the vertical coordinate equal to 1 (galaxy distances having the present values), marked with a dotted line.

Hubble-Lemaitre law will correspond to blueshifts because the wavelengths will shorten when light travels through a contracting space. We don't know what is going to happen when the density becomes again virtually infinite, about 34 billion years after the Big-Bang in this example. If $\Lambda=0$, the age of the universe is about 9 billion years, and the universe will expand forever, but with increasingly slower speed: The expansion is going technically to grind to a halt after an infinite amount of time. The evolution with the current estimate of Λ (a positive value $\Lambda > 0$) shows that the expansion is at first slower than the $\Lambda=0$ case – the predicted age of the universe is about 13.8 billion years – but then increases fast with age, and is expected to continue indefinitely, at an ever-increasing speed.

The future of this universe is interesting, if not scary. With an increasing expansion rate, the local gravity will eventually no longer be able to keep together stars within galaxies. At some stage – but this will happen in tens of billions of years from now – the Milky Way will be destroyed, planets will be detached from their stars and ultimately even stars and planets will be ripped apart by the increasingly fast expansion of space. In an unimaginably far future even molecules, atoms, and nuclei will be torn apart, despite their small dimensions, because of the enormous speed of expansion of space.

Let's continue to focus on the model of our universe, the curve labelled $\Lambda > 0$ in Fig. 2.7. The shape of the curve changes very clearly with increasing age because the density of the various components of the universe changes with time. Upon expansion, the density of matter decreases, with the density of fast particles like photons getting smaller at a rate faster than the density of massive, slow particles. On the other hand, the energy density associated with the cosmological constant does not change. It is like energy inherent to space itself, and as new space is 'created' through the expansion, this energy density remains constant and eventually – about 8 billion years after the Big-Bang – dominates the matter/energy budget of the universe. At this point, the acceleration of the expansion of the universe sets in.

To understand why a cosmological constant accelerates the expansion of space, we may consider the following example. After we have filled with a gas of ordinary matter (including photons) a cylinder with a movable piston on one end, we place it in a chamber from which all air has been removed. The gas will try to increase its volume by pushing on the surface of the piston that slowly moves out, because a gas of ordinary matter has a positive pressure. Matter with positive pressure slows down the expansion of the universe, according to the equations of the Friedmann-Lemaitre-Robertson-Walker models. A hypothetical gas with the same properties of a cosmological constant would instead try to pull the piston into the cylinder and decrease its volume, because it has negative pressure; hence the cosmological constant causes an acceleration of the expansion of the universe.

We do not have (yet) a theory that explains the nature of the cosmological constant, therefore we cannot be sure that it will stay constant forever. Current measurements simply tell us that it has been a constant during the past evolution of the universe but, as far as we know, it could start varying at some point in the future, hence the history of the future expansion of the universe is at this stage somewhat speculative.

2.3.1 Nucleosynthesis and the Origin of the CMB Radiation

Let's now briefly sketch the main events during the early evolution of the universe. Around 10^{-43} seconds after the Big-Bang (we cannot say anything about what happened before this time) the expanding universe was much denser and hotter than today, with temperatures on the order of 10^{32} K. By 10^{-11} seconds after the Big-Bang, the universe had cooled down to temperatures on the order of 10^{15} K, and all building blocks of the visible matter (quarks and leptons – see the appendix for more details) were formed. The universe at this stage was likely made of quark-antiquark, lepton-antilepton pairs plus photons and dark matter. The high energy of the photons allowed the production of particle-antiparticle pairs when their energy was equal to at least twice the mass of a given particle species, multiplied by the square of c (see the appendix). These particle-antiparticle pairs eventually annihilated giving back photons, but additional pairs were then produced, to replenish their number. Because of these processes photons and other particles of visible matter interacted with each other so often, that temperature differences among all these components (if any) were erased almost immediately. It is like having bodies at different temperatures: When brought into contact with each other, they will all reach the same temperature (see the appendix). With particles so tightly 'coupled' by frequent interactions, the radiation (the photons) behaves like a blackbody, peaked at very short wavelengths in this early universe, because of the high temperatures.

When the energy of the photons decreased (due to the expansion of the universe) below the threshold for the production of a species of particle-antiparticle, these pairs annihilated without being replaced. Somehow, some excess matter survived: Our existence is proof that the laws of nature treat matter and antimatter slightly differently, but it is not yet clear the mechanism at work in the early universe.

About 100 seconds after the Big-Bang the temperatures were already down to 10^9 K, protons and free neutrons[14] had been formed starting from combinations of quarks, and nuclear reactions started combining protons and neutrons to produce nuclei of more complex chemical elements (more on nuclear reactions in the next chapter). After just a few minutes the reactions ceased because of the fast expansion and decrease of temperature. At this point, 75% of the mass of the visible matter was hydrogen (just one proton), plus about 25% of helium (two protons and two neutrons), with traces of deuterium (^{2}H, an isotope of hydrogen with one proton and one neutron, see the appendix), ^{3}He (an isotope of helium with two protons and one neutron), and lithium (three protons and four neutrons). The values of these abundances depend on the present density of the visible matter in the universe, which determines its density at the time of this 'cosmological nucleosynthesis'.

For the next 380,000 years of expansion, the visible matter remained still tightly coupled (meaning frequent interactions) to photons through electron scattering. In this process photons are scattered by electromagnetic interactions with free electrons, changing randomly the direction of their path at each interaction: Due to this frequent scattering that randomizes the direction of their motion, photons cannot travel appreciable distances.

The temperature of photons and matter continues to decrease steadily, and at this stage we may now ask ourselves why does expansion cause a decrease of temperature? It is possible to give a qualitative explanation as follows. For the photons, the expansion of space stretches their wavelengths as previously discussed to explain the origin of galaxy redshifts, and longer wavelengths mean photons corresponding to a blackbody with decreasing temperature. In case of matter particles, like for example protons, the temperature depends on their velocity: Slower velocity means lower temperature (see the appendix). We may consider a system of particles with velocities of 100 m/s, measured in a universe that is static: If expansion starts abruptly at a speed of 10 m/s, the velocities of these particles relative to space become equal to only 90 m/s, and correspondingly their temperature decreases. This decrease of temperature with expansion explains why

---

[14]Neutrons are unstable when they are not in an atomic nucleus, and were transforming ('decaying') into protons, electrons, and antineutrinos. Given a sample of free neutrons, it takes about 10 minutes for half of them to decay (see the appendix).

the temperature of the universe close to the Big-Bang is predicted to have reached enormously high values.

Around 380,000 years after the Big-Bang the temperature of the visible matter and the photons has decreased to about 4000 K, cold enough for the electrostatic interactions to combine the free electrons with the hydrogen nuclei, and form electrically neutral hydrogen atoms. This process is known as 'recombination' (although actually electrons and nuclei were combining for the first time in the history of the universe) and given that photon scattering on free electrons is no longer efficient, visible matter and radiation decouple, their temperatures starting to evolve separately.

The temperature of the photons originated during the Big-Bang continues to drop with the expansion, but they keep behaving like a blackbody. Since photons do no longer interact with matter, they can now travel undisturbed through space. In the Big-Bang cosmological model these are the photons that make the radiation observed today as CMB. The CMB is therefore like the echo of a distant time, providing us with a snapshot of the universe only a few hundred thousand years after the Big-Bang. During their 13.5 billion years' travel through expanding space, the wavelengths of these primordial photons have been heavily redshifted, corresponding today to a blackbody with much lower temperature compared to the time of recombination, on the order of $\sim 3\,\mathrm{K}$, as the CMB. These photons are also expected to be distributed isotropically in space like the CMB, because of the cosmological principle. In fact, as early as 1948 Ralph Alpher and Robert Herman published a paper noticing that, if the universe began hot and dense as the Big-Bang model required, it must be currently permeated by photons with a blackbody energy distribution, at a temperature estimated at the time to be around 5 K.

The tiny temperature fluctuations observed in the CMB are required by the presence of galaxies and clusters of galaxies today. If the universe was perfectly isotropic and homogeneous at recombination, it would have been impossible for the gravity to start forming the structures we see in the cosmos. Some density inhomogeneities had to exist, with regions denser than the background starting to contract and get denser still, thus inducing a growth of the initial perturbations. Between 1970 and 1972 Edward Harrison, Jim Peebles (who was awarded the 2019 Nobel Prize also for his work on the CMB), Rashid Sunyaev, Jer Tsang Yu, Yakov Zeldovich, predicted the existence of a signature of these inhomogeneities in the CMB as tiny temperature fluctuations, with denser regions causing higher temperatures. Only density fluctuations of the visible matter couple to the photons and cause the observed CMB fluctuations, because the more abundant dark matter ignores them. The observed temperature fluctuations of the CMB can then be explained by assuming that the density fluctuations were somehow triggered at the same time everywhere in space.

It is now possible to sketch briefly how the values of the various matter/energy densities and Hubble constant are estimated. The power spectrum of the CMB constrains mainly Ω, hence the geometry of space, plus the density of visible matter, but there is also a dependence on the expansion history of the universe. Studies of the velocities and motions of galaxies provide Ω_{DM}, whilst measurements of distances and redshifts of galaxies allow us to determine both H_0 and how the expansion rate of the universe has changed in the past, if we push our observations up to $z=1$ (that corresponds to a distance of about 10 billion light-years) and beyond. Today we can observe the slope of the relation between z and distance D changing with increasing redshift, due to the changes in the expansion rate of space, that depend on the values of the matter/energy densities, including Ω_Λ[15]. Finally, an additional constraint on the expansion history of the universe comes from studies of the distribution in space of galaxies (the visible matter) at different redshifts. These studies show that it is slightly more likely to have galaxies separated by a certain value of the distance, and that this 'preferred' value has evolved over time (with redshift) as the universe has expanded. The existence of this preferred distance is a consequence of the power spectrum of the CMB temperature fluctuations that I briefly mentioned before, and its variation with redshift is only due the expansion history of the universe. It is the joint analysis of all these constraints has led to the accepted values of H_0, Ω_γ, Ω_ν, Ω_M, Ω_{DM}, and Ω_Λ listed before.

2.3.2 Inflation

Everybody is familiar with the concept of horizon in everyday life: Horizon is the line at the farthest place that you can see. A similar 'line', so to speak, does exist in the universe. In this case, the horizon is a measure of the farthest distance from which one could possibly

---

[15] Saul Perlmutter, Adam Riess, and Brian Schmidt were awarded in 2011 the Nobel prize for the discovery of the accelerating expansion of the universe, interpreted in terms of a positive cosmological constant.

retrieve information through the exchange of photons, that travel at the fastest possible speed. I am referring here to what is technically called particle horizon: There are other types of horizons in cosmology, but they are not relevant to our purposes.

At any stage during the evolution of the universe, there is only a finite amount of time elapsed since the Big-Bang, and a limited distance – the horizon – that photons can travel. The distance to the horizon today is 46 billion light-years, the edge of the observable universe, the region of space from which we can get information, even though this information comes from the past.

It might be surprising that the horizon is beyond 13.5 billion light-years, for light could travel at most for about 13.5 billion years, the CMB photons 'released' at recombination. The reason is that space has been expanding since the Big-Bang. As discussed before, photons can start moving freely across the cosmos 380,000 years after the Big-Bang, and in their travel, they have to 'fight' against the expansion of the universe, which keeps stretching space and increasing distances. Those that reach us as CMB, have started their travel from points whose distance is 'today' equal to 46 billion light-years. But we cannot observe what lies 'at this time' at the horizon, because an object that is now at a distance of 46 billion light-years is receding from us at a speed higher than the speed of light[16].

Let's now consider the universe at the time when photons started to travel unimpeded through space. The size of the horizon for any point in space was on the order of 200 Kpc, about 650,000 light-years. This means that only points within a 200 Kpc radius could have exchanged particles between the Big-Bang and the epoch of recombination. If we now 'expand' the size of this horizon, matching the actual global expansion of the universe, we find that it would cover only a very small patch in the sky. Points outside this patch could not have 'communicated' by the time of recombination, but the CMB looks today the same in any direction across the whole sky. As a consequence, there isn't a process that could have produced this uniformity, and the only solution is to accept that the observed uniformity of the CMB over the whole sky is an 'initial condition', set up at the Big-Bang.

---

[16]I recall that this recession velocity is due to the expansion of space, not an actual faster-than-light motion of the galaxies.

The same issue pops up when we consider the geometry of space. Observations point to $\Omega=1$ today, hence space is flat. The Friedmann-Lemaitre-Robertson-Walker models tell us that $\Omega=1$ today means a value different from 1 by less than about 10^{-13}, one second after the Big-Bang. The universe seems to have had $\Omega=1$ imprinted at birth, another initial condition.

Yet another question relates to the presence of the tiny temperature fluctuations of the CMB. They are required by the existence of stars, galaxies and clusters of galaxies, but how did they appear if the universe was supposed to be homogeneous and isotropic? How could they all be triggered at the same time? Is this another initial condition?

Faced with these fundamental questions about the nature of the universe at the Big-Bang, we had basically two options. One option is to accept these initial conditions as given and requiring no further explanation. Another option is to seek a physical mechanism that explains how these initial conditions have arisen. This can be seen perhaps as a philosophical decision to take, and science went for the second option.

During the late 1970s and early 1980s, thanks to the works by Allan Guth, Katsuhiko Sato, Andrei Linde, Andreas Albrecht, Demosthenes Kazanas, Paul Steinhardt, and Alexei Starobinsky, the so-called inflationary paradigm took shape and has become widely accepted, even though – and this is important to point out – it is based on physics that cannot be tested by experiments here in our labs[17]. The idea is conceptually simple. About 10^{-35} seconds after the Big-Bang, a tiny patch of space, possibly of subatomic size, got filled with a new form of energy that acted like an enormously large cosmological constant (but not the same cosmological constant that enters Ω_Λ) triggering a very short period of accelerated expansion, much faster than the speed of light (hence the name inflation). The expansion lasted until about 10^{-32} seconds after the Big-Bang, by which time the size of this patch – the patch of the universe we inhabit – increased by at least 10^{30} times, an impossibly large number.

The uniformity of the CMB can then be explained by this inflation, because distant points in the sky at the same CMB temperature today were actually within their corresponding horizons before the inflation. Therefore they could have equalized temperatures, like a hot and a cold

---

[17]Very recently, Steinhardt has turned against this idea, even saying that inflationary cosmology cannot be evaluated using the scientific method. On the other hand, Guth and Linde are convinced that inflation is on a stronger footing than ever before.

body reach a common temperature when brought into contact. Also, the flatness of space does no longer need to be an initial condition in this scenario. Imagine living in a two-dimensional world, like being squashed on the surface of a tennis ball, that is curved and has the geometry of a sphere. If we now expand the ball to the size of the Sun, the surface of our world will look flat as far as we can see, even though it is a sphere on an enormously (compared to our dimensions) large scale. Inflation stretches any initial curvature of the three-dimensional space to virtual flatness.

Finally, inflation can potentially account for the CMB temperature fluctuations. On the short timescales and sizes involved (the subatomic initial size of the inflationary patch) the amount of inflationary energy is not well defined, in the sense that it is subject to random fluctuations according to the laws of quantum mechanics. These fluctuations on subatomic scales are stretched during inflation into cosmic-sized regions with different amounts of energy. The frantic expansion of space is envisaged to stop when the inflationary energy transforms into matter and radiation, but different regions will stop expanding at slightly different times, depending on the local density of this energy, producing a pattern of slightly different densities and temperatures of the universe. This pattern is then imprinted in the CMB temperature at recombination. Moreover, as just mentioned, the inflationary energy is supposed to transform into essentially all matter and radiation we find in the universe today. The Big-Bang might then be identified with this moment, whilst the origin of cosmic time is set at the fictitious (because we do not have a theory for this regime) instant of infinite density that predates the onset of inflation.

Although this view of the origin of the universe is widely accepted, it is important to point out that we do not know what this inflationary energy may be, and some kind of 'tuning' of its properties is required to match the observed pattern of CMB fluctuations. .

2.4 THE FORMATION OF GALAXIES

The model for the formation and evolution of the universe I have just summarized, is also called the 'ΛCDM' cosmological model, because the matter/energy density is dominated today by the cold dark matter (CDM) and the dark energy/cosmological constant (Λ). With this background, a crucial question to be answered is the following: Is it possible to explain the wealth of structures we observe, galaxies of different types, galaxy groups and clusters, their space distribution, and evolution with time, within the framework of the ΛCDM model?

There is today general consensus on the following broad picture. The density fluctuations of the visible matter did not grow much until decoupling, because they were kept 'in control' by the interaction with photons (in fact the temperature fluctuations of the CMB temperature are extremely small). On the other hand, the density fluctuations of the dark matter could grow, because they weren't dampened by the interaction with photons. These dark matter condensations produced stable structures called 'haloes', whereby random motions of the dark matter particles counterbalanced their self-gravity. These haloes merged with time, producing increasingly massive structures.

After recombination also the visible matter started to travel freely and was dragged along by the CDM that by now dominated gravity. The dark matter haloes began accreting visible matter so that galaxies are the product of the evolution of visible matter nested inside larger CDM haloes. Processes like merging between haloes of similar mass, accretion of much less massive haloes and diffuse gas, destruction of small haloes when falling onto larger ones, drive the growth of the cosmic structures over time. A very important result is that galaxy morphology may be a transient phenomenon and the various types of galaxies we observe today reflect the variety of accretion histories.

Modelling these processes is very complex and subject to sizable uncertainties. The 'easiest' part is to follow the evolution of dark matter because it is shaped only by gravity. The behaviour of the visible matter is much more complicated to follow, for it requires a detailed description of its accretion onto CDM haloes, heating and cooling (electromagnetic interactions) plus the conversion of gas into stars (the star formation processes), and the chemical evolution driven by the evolution of stars, which inject into the intergalactic medium gas enriched in elements produced by nuclear reactions (see next chapters). Several of these processes are still poorly understood in a galactic context, and we need to employ phenomenological prescriptions which contain free parameters to be somehow tuned.

A powerful way to provide firm, independent constraints on galaxy formation models, and particularly on the evolution of the visible matter, is to determine the ages of stars. They paint a picture of the star formation history of the various galactic stellar populations which, paired with determinations of their chemical composition (and velocities, when available), allow us to trace the history of the accreted gas, its efficiency in forming stars, the timescales of its chemical enrichment due to stellar evolution.

The importance of the constraints posed by stellar ages is perfectly highlighted by the following example. During the mid-1990s models for the formation and evolution of elliptical galaxies predicted that more massive ellipticals host younger stellar populations than less massive ones, and that ellipticals in lower mass CDM haloes are typically younger than ellipticals in more massive haloes. These predictions were disproven by determinations of the ages of stars in elliptical galaxies, forcing a reexamination and improvements of the physics of galaxy formation models.

In the next chapters we will see how it is possible to determine the ages of stars. But first, we need to learn about stars' inner workings.

GLOSSARY

Cosmic time: Time measured by clocks that are stationary relative to the expanding space, set to zero at the hypothetical initial instant of infinite density.

Cosmological nucleosynthesis: Production of chemical elements a few minutes after the Big-Bang.

Cosmological principle: Cornerstone of the current cosmological model. The principle states that any view of the universe, at a given cosmic time, is independent of the direction in which the observer looks, and of the observer location.

Dark matter: Unknown type of matter that does not absorb, reflect, or emit light. It makes about 27% of the universe.

Dark energy: Unknown type of energy that makes approximately 68% of the Universe. It appears (so far) to be in the form of a 'cosmological constant', meaning that its density does not change with the expansion of the universe.

Gravitational lensing: Distortion and bending of light from distant galaxies, caused by the gravitational pull of masses located between source and observer. Strong gravitational lensing can even result in multiple images of the same galaxy.

Inflation: Very short period of rapid expansion of space (by a factor of at least about 10^{30}), between about 10^{-35} and 10^{-32} seconds after the Big Bang.

Kelvin temperature scale: The Kelvin (K) is the primary unit of temperature measurement in physics. It has the same magnitude of the degree Celsius (C), but a different zero point. A temperature of zero Kelvin corresponds to -273.15 on the Celsius temperature scale.

Lookback time: The time elapsed between the detection of the light from an astronomical object here on Earth, and when it was originally emitted by the source.

Parsec: Unit of distance corresponding to 3.26 light-years.

Recombination: Event in the history of the universe corresponding to the first time when free electrons and protons combined to form neutral hydrogen atoms (about 380,000 years after the Big-Bang).

FURTHER READING

S.F. Green and M.H. Jones. *An Introduction to Galaxies and Cosmology*. Cambridge University Press, 2015

A. Guth. *The Inflationary Universe: The Quest for a New Theory of Cosmic Origins*. Basic Books, 1998

S. Singh. *Big Bang: The Origin of the Universe*. Harper-Collins, 2005

R.S. Somerville and R. Davé. Physical Models of Galaxy Formation in a Cosmological Framework. *Annual Review of Astronomy and Astrophysics*, 53:51, 2015

S. Weinberg. *The First Three Minutes: A Modern View Of The Origin Of The Universe*. Basic Books, 1993

CHAPTER 3

Set the Controls for the Heart of the Stars

Columbia and Wrightsville are two small towns in Pennsylvania, United States, on the opposite banks of the Susquehanna River, connected by the Veterans Memorial Bridge, the world's longest concrete multiple-arch bridge, opened in 1930. If you want to cross the bridge from Columbia, follow the Lincoln Highway that will lead you to the bridge right after crossing Rotary Park. For two kilometres you then drive over the waters of the Susquehanna, a name derived from a 'Len'api' term meaning 'oyster river', because oyster beds were once widespread in the Chesapeake Bay, near the mouth of the river. At the end of the bridge in Wrightsville, these days you'll find a pizza restaurant and a hairstyling salon at the opposite sides of the road, followed by a bicycle shop, a food market, and then another pizza restaurant.

In the last years these two towns and the Veterans Memorial Bridge have witnessed something remarkable: Invasions of swarming mayflies. Mayflies are small aquatic insects, that spend a few years as larvae (at most a few centimetres long) feeding on algae, plants, and organic matter on riverbeds. They emerge as adults from spring to autumn (not necessarily in May as suggested by their name) as delicate-looking insects with one or two pairs of triangular wings, to die typically within 24 hours or less. Often, all the mayflies in a population mature at once (a 'hatch'), with millions of insects rising up from rivers – like the Susquehanna – or lakes in thick clouds, to then going to land on houses, stores, cars, but also people's arms, legs, hair... The local news website *Lancaster online* was reporting in June 2015 the words of the Columbia fire chief, who said that it was like a blizzard of mayflies, with zero visibility for a quarter-mile across the bridge. The firemen had to close their eyes, swat the mayflies away because they were getting into their mouths, and continued to have the feeling of bugs crawling on them even when they were back to the fire-station.

Although the lifespan of these insects is a minuscule fraction of a human lifetime, some inquisitive mayfly might be attempting to understand the nature of these giant lifeforms trying to slap them. Finding an answer to this question would be a really daunting task: One lifetime of observations (typically 24 hours) would tell the mayflies that these beings do not change their appearance at all. Granted, they might be moving at some point, then stop, then move again, but their size, the dimensions of their limbs do not seem to change. Are they immortal? Are they born as we see them and stay like this forever? That would be the simplest explanation... but there are many of them, in different sizes, slightly different shapes and characteristics, and maybe – some clever mayfly might be speculating – we are seeing these creatures at various stages of their lives. Different sizes may reflect different milestones during their life-cycle. This interpretation has the potential to give us clues about how they develop, but to decipher these clues we badly need to know the basic principles driving the development of these giants.

Trying to understand how stars work and evolve in time, is a task similar to what our scientist mayfly might be attempting: If anything, the difference between human and stellar evolutionary timescales is bigger than between mayflies and us. For this reason, the development of the theory of stellar structure and evolution can be rightly considered one of the crowning achievements of twentieth-century physics, and science in general, and a perfect case study for the need to bring together scientists of different branches of physics, to advance our understanding of nature.

3.1 EARLY IDEAS

The first attempts to build a scientific theory of what stars are made of and how they work, had to wait until the laws of thermodynamics and the concept of conservation of energy were established. Before that time, even a renowned scientists like William Herschel – the discoverer of Uranus and the first researcher to envisage the presence of a disk of stars in the Milky Way – could propose that the Sun was actually inhabited. In 1795 he suggested that the Sun has a solid surface, surrounded by a hot, glowing atmosphere. The dark sunspots were holes in the atmosphere which provided a view of the solid nucleus below, inhabited by *beings whose organs are adapted to the peculiar circumstances of that vast globe*. The Sun was considered to be like a planet, and given that stars were imagined as objects of the same nature of the Sun, Herschel concluded that all stars *themselves are primary planets*. These ideas weren't at all unusual, as other scientists of the time imagined volcanoes and mountains on the solid surface of our star. Also in those years at the end of the eighteenth century, thanks to Emanuel Swedenborg, Immanuel Kant, and Pierre-Simon Laplace, the idea that stars form from gas clouds collapsing under their own gravity started to take root.

At the beginning of the nineteenth century, there was scepticism about the possibility to build a scientific theory about the nature of stars, as exemplified by the writings of the famous philosopher August Comte. According to Comte and his 'positivism', every scientific theory must be based upon observed facts, and in 1835 he wrote that it is possible to study the shape, distances, sizes, and motions of stars, but we will never know how to find out their chemical composition and average temperatures. This statement was soon to be disproven, thanks to the observations of the solar spectrum by William Wollaston in 1802, and 12 years later by Joseph Fraunhofer, that paved the way to the discovery of the chemical composition of the outer layers of stars.

Taking the spectrum of the light received from the Sun or any other star means to shine the beam of light through a 'spectroscope', an instrument that bends the path of light rays of different wavelengths by different angles. The output of this instrument is a strip of light, a narrow band of colours (the 'spectrum', from the Latin word meaning *image*) just as one would see in a rainbow; when moving across the strip we go through light of different wavelengths, that has been deflected at different angles. The red colour comprises light with wavelengths between about 625 and 740 nm (nm stands for nanometres, and one nanometre is equal to 10^{-9} metres[1]) and, moving through orange, yellow, green, cyanogen, blue, we reach the violet colour, that covers wavelengths between about 380 and 450 nm. Isaac Newton was the first to do this, shining sunlight through a prism, which came out separated into all the colours of the rainbow, like on the iconic cover of a very famous album of rock music. The same effect is at play when

---

[1] For comparison, a human hair is typically around 100,000 nm wide.

we see a rainbow in the sky, due to the presence of droplets of water suspended in the atmosphere which, in the presence of sunlight, behave exactly like small prisms.

Wollaston, however, noticed that the apparent continuous strip of light that made the spectrum of the Sun was crossed with several 'dark lines', that he thought were divisions between the colours. Fraunhofer discovered 574 of these lines in the solar spectrum, scattered amongst the colours, and he also realized that the spectra of the other few stars he observed were different from the solar spectrum.

The discovery of all these dark lines was made possible by the development of a new kind of instrument to disperse light. Instead of a prism, a so-called diffraction grating was employed, typically a transparent or opaque plate of material, crossed by several hundreds of very fine lines per centimetre. This grating disperses the light more evenly than a prism, with a regular and very simple relation between the position of the spectral lines and their wavelength, plus the dispersion is greater than when using a prism. This latter property meant that different wavelengths were spread much further apart, facilitating the discovery of these mysterious black lines, which are specific wavelengths where the number of photons in the light beam is much lower (hence they appear dark) than in the neighbouring regions of the spectrum.

It did not take long before scientists realized from laboratory experiments, that each chemical element has its own characteristic spectrum, like an individual fingerprint. These results followed the work of the physicist Gustav Kirchhoff and the chemist Robert Bunsen in the German town of Heidelberg. In the 1860s Kirchhoff formulated the following three rules (see Fig. 3.1):

- A hot solid, liquid, or highly compressed (high density) gas emits a continuous spectrum of light (like a rainbow), without any lines.

- A hot gas under low pressure (low density) gives off a spectrum made of bright lines on a dark background. The number and wavelength of these lines depend on the elements present in the gas (like a barcode that uniquely identifies the chemical species).

- If a continuous spectrum passes through a low-density gas at a lower temperature, this cooler gas generates dark absorption lines, whose number and wavelength depend on the element in the cool gas.

These empirical rules highlighted the equivalence between the lines seen in absorption in stellar spectra and the emission lines seen in the spectrum of sparks and flames, generated by different elements in laboratory experiments. Emission and absorption lines with the same wavelength were produced by the same chemical element.

By comparing the wavelength of the lines in the solar spectrum with emission lines seen in laboratory spectra, Kirchhoff deduced the presence of iron, calcium, magnesium, sodium, and chromium in the outer layers of the Sun. We were finally discovering what the Sun – at least its surface layers – was made of, less than 30 years after Comte's statement, even though there was no theoretical understanding of how the absorption lines were formed.

Figure 3.2 displays an example of modern spectrum of a star similar to the Sun at wavelengths (denoted with λ) between 370 and 740 nm, displayed as brightness (in arbitrary units) versus wavelength. The average trend, showing an increase of the average brightness with λ up to around 500 nm, followed by a gentle decrease, is the continuous spectrum, while the superimposed 'wiggles' are caused by the absorption lines.

The ability to determine – at this stage just qualitatively – the chemical composition of the surface of stars, with the discovery of the same elements we find here on Earth, changed forever our view of the universe. As reported by Agnes Clerke[2] in her 1885 book *A popular history of astronomy during the nineteenth century*, this was one of the groundbreaking developments of nineteenth century astronomy: *That a science of stellar chemistry should not only have become possible, but should already have made material advances, is assuredly one of the most amazing features in the swift progress of knowledge our age has witnessed....... the application of prismatic analysis certified to the presence in the stars of the familiar materials, no less of the earth we tread, than of the human bodies built up out of its dust and circumambient vapours.*

---

[2]Agnes Clerke was a prolific contributor to several periodicals, publishing both historical surveys as well as science books, aimed at explaining to the general public the latest scientific developments. *A popular history of astronomy during the nineteenth century* is probably her most celebrated contribution to astronomy. In 1903 she became the fifth woman to be elected as a member of the Royal Astronomical Society. In 1973, the International Astronomical Union named a crater in her honour, and in 2017, 110 years after her death, the Royal Astronomical Society established the Agnes Clerke Medal, awarded for outstanding research into the history of astronomy or geophysics.

SPECTRUM

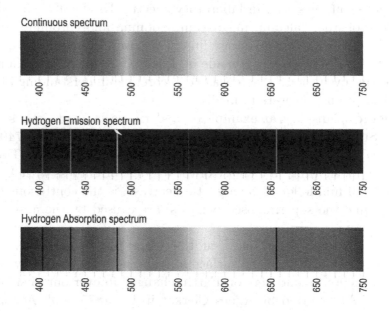

Figure 3.1 Examples of continuous, emission, and absorption spectrum, the three types of spectra according to Kirchhoff rules. The emission and absorption spectra are for hydrogen gas. The horizontal axis displays wavelengths in nanometres (Designua/shutterstock).

The application of the results of spectroscopy obtained from laboratory experiments to the study of stars marks also the beginning of a new type of astronomy, first named astrophysics by Johann Zöllner in 1865. This new astronomy was at first dismissed: Famous astronomers of the time, like the seventh Astronomer Royal George Airy, director of Greenwich Royal Observatory between 1835 and 1881, shared the same views expressed by the following comment made around 1840 by

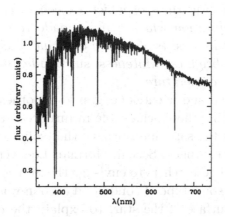

Figure 3.2 Example of absorption spectrum for a star similar to the Sun, displayed as measured brightness (in arbitrary units) as a function of wavelength [110].

Friedrich Bessel[3], as reported in Edward Maunder book about the history of Greenwich Observatory, published in 1900: *What astronomy is expected to accomplish is evidently at all times the same. It may lay down rules by which the movements of the celestial bodies, as they appear to us upon the earth, can be computed. All else which we may learn respecting these bodies, as, for example, their appearance, and the character of their surfaces, is, indeed, not undeserving of attention, but possesses no proper astronomical interest....... To learn so perfectly the motions of the celestial bodies that for any specified time an accurate computation of these can be given–that was, and is, the problem which astronomy has to solve.*

Despite this scepticism, in the decade between 1860 and 1870 studies of the solar spectrum using Kirchhoff laws advanced our ideas about the nature of the Sun (and stars in general). The Sun was envisaged first as an incandescent liquid by Kirchhoff, Heinrich Magnus and Kelvin. Kirchhoff wrote in 1861 in his *Researches on the solar spectrum and the spectra of the chemical elements* [4] that *in order to explain the occurrence of the dark lines in the solar spectrum, we must assume that the solar atmosphere incloses a luminous nucleus, producing a contin-*

---

[3]Bessel made the first successful measurement of the parallax of a star in 1838 (see the appendix).

[4]English translation by H.E. Roscoe, for Macmillan and Co., Cambridge, 1862.

uous spectrum, the brightness of which exceeds a certain limit. The most probable supposition which can be made respecting the Sun's constitution is that it consists of a solid or liquid nucleus, heated to a temperature of the brightest whiteness, surrounded by an atmosphere of somewhat lower temperature.

Ideas evolved fast, and scientists began to hypothesize for the Sun a temperature exceeding those achievable in furnaces: It was now inconceivable to think of the solar interior as either solid or liquid. Herbert Spencer, Hervé Faye, Angelo Secchi, Norman Lockyer, Homer Lane[5], were among the first researchers to envisage the Sun as a gas sphere, although graphite or some type of condensed carbon was still considered to be part of the surface of the star, to explain the continuous spectrum in the background of the observed absorption lines. This latter hypothesis was related to the fact that the continuous spectrum was beginning to be considered equivalent to what we call a blackbody, like the spectrum observed from a tiny hole in the side of an oven or a furnace. The colour of the light inside a furnace depends on its temperature, becoming bluer with increasing temperature (the Wien law). On Earth, only graphite and soot were known to produce this kind of spectrum.

The paradigm-changing nature of these new views on the Sun should not be underestimated. Clerke comments about Faye's description of the structure of the Sun presented first at the Paris Academy of Sciences in 1865, summarize well the nature of this revolution: *The essence of the change may be conveyed in a single sentence. The Sun was thenceforth regarded, not as a mere heated body, or – still more remotely from the truth – as a cool body unaccountably spun round with a cocoon of fire, but as a vast heat-radiating machine.*

More crucial theoretical developments followed in the last decades of the nineteenth century. Rudolf Clausius delivered in 1870 a lecture to the Association for Natural and Medical Sciences of the Lower Rhine, in Germany, where he presented the 'virial theorem', an equation that

---

[5]Homer Lane is remembered for his 1870 paper titled *On the theoretical temperature of the Sun*, the first attempt at a mathematical description of the interior of a star. After graduating from Yale he worked first at the Patent Office in Washington D.C., then as a patent lawyer, inventor, sellers of oil lands, followed by work for the U.S. Coast Survey. In his spare time, he developed his model for the Sun and was elected member of the National Academy of Science in 1872, but remained a little-known figure in the American scientific community. The crater Lane on the Moon is named after him.

relates the average value over time of the total kinetic energy (the energy associated with the motions of the particles) of a stable system made of particles bound by attractive forces, with that of its total potential energy. This result has been since applied widely in astrophysics, and had a crucial impact on our understanding of how stars work, as we are going to see right now.

The fact that the Sun doesn't change its radius over timescales of human generations has an important consequence: The Sun (and by extension all the stars) is in hydrostatic equilibrium, meaning that at any point the force of gravity, that would tend to make the gas collapse towards the centre, is balanced by the pressure of the gas, whose effect is to push outwards the various gas layers. This can be easily demonstrated by calculating how long it would take a blob of gas at the surface of the Sun to reach the centre, due to the gravitational force it is subject to. It is a simple calculation that requires only the knowledge of the mass and radius of the Sun, and the value of the gravitational constant: The result is a time on the order of 10 minutes.

Hydrostatic equilibrium was indeed the first assumption made by Augustus Ritter when he applied the virial theorem (valid for stable systems) to stars. The second assumption involved the nature of the gas in the Sun, meaning how pressure, density, and temperature are related, the so-called equation of state (see the appendix). The simplest assumption is that of a 'perfect gas', a gas composed of many randomly moving particles, whose sizes are negligible compared to the size of the system in which they are contained (the whole Sun in this case), and whose only interactions are perfectly elastic collisions. When two particles experience such a collision, the total kinetic energy of the two bodies remains the same, there is no conversion into heat or potential energy, roughly like a ball at a billiard table that hits another ball, and the two objects bounce apart. Actually, the gas in the Earth's atmosphere is to a good approximation a perfect gas, and the higher the temperature the more realistic the approximation.

For a sphere (like the Sun) made of a perfect gas in hydrostatic equilibrium, the virial theorem provides a surprising result. As the Sun loses energy from its surface because it is hotter than the surrounding space, it needs to slowly shrink to rebalance gravity and gas pressure. Half of the gravitational energy released by the contraction replaces the energy lost from the surface, the other half is converted into thermal energy and raises the temperature of the gas, so that the pressure is able to balance the greater gravitational forces of its more compact state. In a nutshell: The star loses energy, and to restore hydrostatic equilibrium it contracts and at the same time heats up. This result by Ritter was obtained independently also by Homer Lane[6] and contradicted the general belief that due to gravity, stars in hydrostatic equilibrium slowly shrink and cool due to the loss of energy from the surface, as envisaged by Kelvin and Helmholtz only a couple of decades earlier (see the discussion in chapter 1). Assuming that the solar luminosity has not changed with time, Ritter concluded that due to the slow contraction required by the virial theorem to produce the energy given off by the Sun, its radius was 215 times the present value about 5.5 million years ago – which is about the size of the Earth orbit. As a result, the age of the Earth couldn't be larger than this value. Ritter also proposed a theory of stellar evolution whereby a star begins its life as a diffuse mass of gas which contracts and heats. At some point, the perfect gas approximation ceases to be valid and after a brief period of fixed temperature, the star enters a cooling phase. A qualitatively similar idea about the evolution of stars was proposed in 1887 by Normal Lockyer[7]. At this time gravity was the only known energy source able to power the Sun.

The works by Ritter, Lane, and Robert Emden between 1870 and 1907 provided the mathematical formalism to calculate the first models of the structure of stars, that was used until the advent of computers in the 1950s. This is the so-called Lane-Emden equation, a differential equation (see the next chapter for a simple explanation of what a differential equation is) derived under the assumption that stars are spheres of gas in hydrostatic equilibrium, and that the gas pressure P varies as a power of the density ρ across the star, namely $P = K\rho^n$, where K and n are two constants (two fixed numbers). These early models also assumed that stars were convective, meaning that the transfer of heat across the star was caused by the movement of the gas between areas of different temperatures, like the blobs we see in a pot of boiling water on a stove, continuously cycling from the bottom (close to the source of heat) to the cooler surface. Assumptions about the transfer of heat

---

[6]Lane reached the same conclusions as Ritter very likely around 1868, as reported by his friends in Washington. He never published this result – sometimes called Lane law – that was however attributed to him as early as 1872. Ironically, he is known more for this unpublished result than for his 1870 paper, that presented the first mathematical model of the Sun.

[7]As early as 1865 Zöllner thought that the different colours of stars corresponded to different degrees of cooling during their evolution.

are indeed necessary because energy is continuously flowing from the hot stellar surface to the colder surrounding space, and this cannot last for too long unless the heat content of the surface is replenished from below. It seemed all too natural to identify convection as the culprit, and not just because of the example of the pot of boiling water heated from below: If on a hot day we squat down on an empty road and look at distant objects, we'll notice that their image trembles, with the dreamy quality of a mirage. This is due to the convective movement of air bubbles, cycling between the hot tarmac heated by the Sun, and the overlaying cooler gas layers.

The postulation of convective stars fixes the value of the constant n, but not K. The final step was then to also specify the mass M and the surface radius R of the model. In this way, it is possible to solve the Lane-Emden equation and calculate how pressure, density, temperature, and mass vary from the centre to the surface of the model. These solutions showed that to stay in hydrostatic equilibrium, the cores of stars must be denser and hotter than the surface; given the known values of mass and radius of the Sun, it was soon realized that the temperature at its centre has to be on the order of millions Kelvin, while the surface temperature is on the order of thousands Kelvin.

I close this brief discussion on the Lane-Emden equation, which has played a crucial role in the development of the theory of stellar evolution, to highlight a different way to find solutions, that explains the results discussed in the next sections. If the constant K is also known, we only need to specify M, and the equation can be solved. In this way, the radius of the models is an output of the calculations. With the values of radius and temperature at the surface, the model total luminosity can also be predicted, by employing the recently discovered Stefan-Boltzmann law for blackbodies (see the next chapter). All stellar model calculations in the next sections are based on assumptions that fixed the values of K (and n), so that radius and luminosity were also outputs of the computations.

3.1.1 Stellar Spectra

In parallel with these early developments about the stellar structure, at the end of the nineteenth century Edward Pickering and Willamina Fleming studied thousands of stellar spectra, trying to find common patterns that might eventually lead to a deeper understanding of the properties of the parent stars. They organized the spectra into several types labelled with letters from A to Q, according to the strength (how prominent they are) of hydrogen absorption lines[8]. However, other absorption lines did not follow this continuous pattern, and in 1901 Annie Jump Cannon changed the classification to obtain a continuous sequence of spectral types according to the colour of the star – a measure of its surface temperature. The order of the types was reshuffled to fit the colour progression, some of them were merged, and the final sequence went from the hottest type O (surface temperatures above 30000 K) through B, A, F, G, K, to the coldest type M (temperatures below about 3500 K). Each type was then divided into ten subclasses, from 0 to 9: As an example, spectra whose characteristics were halfway between types G0 and K0 were classified as G5 (the Sun has spectral type G2)[9]

An important conclusion reached by these early studies on stellar spectra was that the same chemical elements as those of our Earth are present throughout the universe. Also, the observed macroscopic differences among spectra of different stars were correctly ascribed to different surface temperatures. However, due to the lack of a theory for the formation of the spectral lines, it was impossible to quantify the abundances of the various elements and assess whether they display star-to-star variations. Intriguingly, in 1914 Walter Adams and Arnold Kohlschütter found that some spectral lines change in strength between luminous and faint stars of known distance. Thanks to this result, they were able to determine a star's intrinsic luminosity directly from the measurement of its spectral lines. This allowed distances (albeit with a large error) to be determined with spectroscopic measurements, a method known as spectroscopic parallax (see the appendix).

The first comprehensive interpretation of spectra by applying the newly discovered atomic physics came in 1925, with the Ph.D. thesis

---

[8]Between 1862 and 1868 Father Angelo Secchi had collected spectra of more than 4000 stars, and developed an earlier classification into four main groups, characterized by different line patterns and strengths.

[9]In terms of line strengths, the spectral type O is characterized by the presence of strong helium lines, that disappear at type B9, whilst hydrogen lines get more prominent when moving to types B and A. At type F the strength of hydrogen lines begins to decline, while lines produced by calcium increase in strength from type A to G. Hydrogen lines become weak beyond type F, whilst lines of the elements heavier than helium get stronger from F to K. Type M sees the appearance of strong lines due to molecular absorption, because in these cold atmospheres various species of molecules can form.

by Cecilia Payne-Gaposchkin[10] titled *Stellar Atmospheres; a Contribution to the Observational Study of High Temperature in the Reversing Layers of Stars* (more on the main results of her thesis in the next section). The presence of the absorption lines can be explained in terms of the energy of the orbits of the electrons around the nuclei, in atoms and molecules. These energy levels are unique to each chemical element, as dictated by the rules of quantum mechanics. Absorption lines occur when an atom (or molecule) –let's call it atom A– in the cool outer layers (the so-called stellar atmosphere) of the star absorbs a photon of wavelength λ_A directed towards the observer, that happens to have an energy exactly equal to the difference between the energy of two electronic energy levels – let's call them levels 1 and 2, where 1 is closer the nucleus. This absorption causes an electron to 'jump' temporarily from level 1 to the more external (higher energy) level 2, but the emission of a photon with the same energy (hence wavelength) of the absorbed one follows shortly after, because the distribution of the electrons around the nucleus is fixed by the temperature/density of the gas and the 'jumping' electron has to move back to its original energy state (level 1). But according to the laws of quantum mechanics this new photon can now be released in any direction, not necessarily towards the observer, so that the absorption process effectively 'extracts' photons of wavelength λ_A from the stellar light measured by the observer, producing the observed spectral line.

To summarize the emerging picture regarding the nature of the stellar spectra, the radiation coming from the hot interior that is released from the surface of a star (called photosphere) has a continuous blackbody spectrum. When these photons cross the cooler atmosphere, spectral absorption lines are formed; their wavelengths are therefore a fingerprint of the chemical makeup of the stars' atmospheres. An absorption line is due to the presence of electrons in certain energy levels (orbits) around the nuclei of a given chemical species, therefore its depth (the number of photons measured at the wavelength of the line) depends on the abundance of that element. However, the temperature also plays a major role, because at a given temperature of the gas, only specific orbits can be populated by electrons, hence only certain

---

[10]Cecilia Payne-Gaposchkin became in 1956 the first woman chair of the Department of Astronomy at Harvard University and during her long career she received several honorary degrees and medals, awarded in recognition of her contributions to science. In 1976, three years before her death, she was the first woman to receive the Henry Norris Russell Prize of the American Astronomical Society.

spectral lines are allowed. The reason is that due to the gas temperature, atoms move around and collide, and the energy of these collisions,
if high enough, can kick some (or all) of the electrons out of their orbits
(first the more external orbits are depopulated because they are less
tightly bound to the nucleus), 'ionising' the atoms. Stars with different
surface temperatures have different temperature distributions in their
surrounding atmospheres, hence the pattern of spectral lines depends
on the temperature of their photosphere.

Finally, there is also the effect of the surface gravity, a measure of
the strength of the gravitational attraction at the surface of a celestial
body; it tells us how fast an object held above the surface of a planet
or the photosphere of a star accelerates when we let it go. On Earth, an
object will accelerate towards the surface increasing its speed by about
9.8 metres per second, for every second of free fall, while on the Sun the
same acceleration is about 30 times higher. A higher surface gravity at
a given effective temperature causes denser stellar atmospheres, with
more collisions among atoms and different strengths of the spectral
lines (this is the explanation for Adams and Kohlschütter results, as
discussed in the appendix).

3.2 THE HERTZSPRUNG-RUSSELL DIAGRAM AND THE DE-VELOPMENT OF STELLAR EVOLUTION THEORIES

Plotting data on a diagram is an essential tool used by scientists in all
fields, to find correlations between measurements of different properties
of a system, like for example height and weight of the human body.
A correlation means that the value of a measured quantity depends
on the value of another one, and can denote the presence of a causal
relationship. This, in turn, provides clues about the processes at work
in the system we are studying. An example is Fig. 2.4 that displayed the
relationship between redshift z and distance D of a sample of galaxies
(higher D leads always to higher z) whose discovery has been crucial
to advance our understanding of the structure and evolution of the
universe.

Between 1905 and 1909 Ejnar Hertzsprung published tables relat-
ing the absolute magnitude (a measure of the intrinsic luminosity) of
nearby stars with known distances, to their spectral type [50]. Although
the spectral types followed an older classification compared to the one
by Cannon that is still used today, Hertzsprung discovered and discussed patterns, relationships between the measured luminosities and

spectra. In 1913 Henry Norris Russell presented at a meeting of the Royal Astronomical Society in London a diagram with the absolute magnitude of stars in the solar neighbourhood, versus their spectral type, following Cannon's classification. Figure 3.3 displays Russell's data in the same type of diagram: The horizontal axis displays spectral types, corresponding to decreasing surface temperatures when moving from type O to type M, while the vertical axis shows the 'absolute magnitude' M_V of the stars in the optical (the wavelength range of the visible light) part of the spectrum. These magnitudes are a proxy for the intrinsic luminosities of stars: Lower values correspond to higher luminosities, as discussed in chapter 6. In this diagram the Sun has a magnitude of about 4.8 and a spectral type G2.

Russell's diagram tells us that luminosity (magnitude) and surface temperature (spectral type) of stars are related, because they display a well-defined pattern, not scattered at random everywhere. Most of them lie along a line that goes from the bottom right to the top left side of the diagram, but a conspicuous number of objects appear also on the right of this sequence, at high luminosities (low values of the magnitude). I have added a line to the diagram, to separate these two groups of stars. Russell called the stars to the right of this line 'giants', and the others 'dwarfs'. The early development of the theory of stellar evolution was rooted in the attempts to explain why stars occupy these specific locations.

This diagram has been named Hertzsprung-Russell diagram (HRD in short) although actually the first plot of this type to be published is not from Russell, as I discovered in June 2001, at a conference titled *Observed Hertzsprung-Russell diagrams and stellar evolution*, held in Coimbra, the former capital of Portugal. I remember the conference very well also because of a curious mishap. It was the day of my talk, scheduled just after the lunch break. At the end of the morning session, I stayed behind in the conference room to discuss with my colleague Constantine Deliyannis and his doctoral student about measurements of lithium abundances in stars, while the other participants were leaving the building to have lunch at the university canteen, next door. The room was on an underground floor, and when we had finished discussing, we went up the stairs to the exit. But in the meantime the caretaker had locked the exit door, assuming that all participants had left for lunch. So we stayed locked up without food and drinks during the whole lunch break. After the caretaker reopened the door, the incredulous reaction from the organizers and several participants

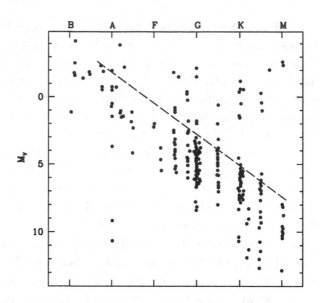

Figure 3.3 Diagram showing the data Russell presented in 1913 and published in 1914 [100]. The dashed line separates the so-called giants (on the right-hand side of the line) from the dwarfs, according to Russell's definitions. This type of diagram was later named Hertzsprung-Russell diagram.

when they unexpectedly found us inside the building was truly amusing. Needless to say, the three of us received countless apologies for the rest of the conference.

My personal discovery of the first HRD, happened on a gloriously sunny and hot Friday, the last day of the conference, when David Valls-Gabaud was giving the concluding remarks at the end of a very interesting week of review talks and contributions. He started his talk by showing the first HRD ever published, that appeared in 1910 in the German journal Astronomische Nachrichten [96]. It was the HRD of stars in the Pleiades, the well-known young open cluster visible to the naked eye in the northern hemisphere, published by Hans Rosenberg as part of his Habilitation thesis (an additional thesis after the PhD, required in Germany to teach in universities).

The horizontal axis of his diagram displayed the difference between the strength of a calcium line and the average of two hydrogen lines, which is also a function of the surface temperature of the star, making the diagram equivalent to the standard HRD. Rosenberg concluded in his paper that there is a tight relationship between brightness and spectral type for stars in the Pleiades (more on this HRD later).

Coming back to Russell's HRD, it was clear that any viable stellar evolution theory needed to explain the distribution of the stars in this diagram. Russell proposed a theory, that echoed the earlier ideas of Ritter and Lockyer [99], under the tacit assumption – actually correct – that the solar neighbourhood is populated by stars of all ages, meaning that new stars have formed continuously since the birth of our galaxy. Assuming that stars emit light like blackbodies, Russell knew that moving towards the left of the diagram (towards higher surface temperatures) at constant luminosity meant also to decrease the radius (see the next chapter), and he envisaged that giant stars were young objects evolving from a diffuse gaseous state towards the dwarf sequence. Following Ritter's and Lane's results, he assumed that giants were contracting and heating up, maintaining roughly a constant luminosity in the process. At some point, due to the high densities reached upon contraction, the perfect gas approximation was no longer valid and stars evolved along the dwarf sequence by contracting and cooling (like Kelvin model for the Sun), eventually reaching a liquid and solid state in their interiors.

Russell summarized his ideas by writing that giant stars represent successive stages in the heating of a body, with redder objects being in earlier evolutionary stages; the dwarf stars instead represent successive stages of the following cooling phase and the reddest objects are in this case the more advanced ones[11]. According to Russell, there was no reasonable alternative to explain the evolution of stars. In this theory the gravitational energy is the stellar energy source, therefore the problem with the age of the Sun compared to requirements from geology and biology was still unsolved. Within about 10 years of its inception, this model was already in trouble, after the works by Arthur Eddington, described in his classic book *The Internal Constitution of the Stars* [34].

As already mentioned, it was generally assumed that stars are convective, the energy released by gravitational contraction being redistributed across their structure by the large-scale motion of gas blobs

---

[11]Because of Russell's ideas about stellar evolution, the spectral types of hot blue stars are still defined as 'early', and those of cold red stars as 'late' types.

that mix the matter – meaning that the chemical composition of the star was necessarily uniform across the star – and transfer energy from one point to another. However, already in 1895 Ralph Sampson proposed that radiation (the photons) could also be an efficient mechanism to redistribute energy within the Sun (radiative transport). Common examples of radiative transport in everyday life are the heat we feel standing close to a lightbulb, or the warming of our body when we sunbathe on a deck chair.

Whenever there is a difference in temperature (like between the centre and the surface of a star) the more energetic photons from the regions of high temperature tend to diffuse towards cooler layers, where they are 'absorbed', releasing their excess energy into these layers. The efficiency of this mechanism depends on the temperature difference between layers (the temperature gradient) and how easily the photons can travel from one place to the other. This ability to travel is quantified by the so-called opacity of the stellar matter, which depends on the chemical composition, temperature, and density. The higher the opacity, the shorter the mean length travelled by the photons, the less efficient the transport of energy across the star. In 1906 Karl Schwarzschild, while working on models of the external layers of the Sun, found a simple criterion to assess whether either radiation or convection are efficient at redistributing energy[12]. Convection sets in when photons are unable to efficiently redistribute the energy within the star because either the opacity of the gas is too high, or because there is too much energy to transport from one point to another (that's why eventually convection sets in a pot full of water when it's heated long enough on a stove).

Schwarzschild found – incorrectly, as we now know – that convection is never efficient in the outer layers of the Sun, under the erroneous assumption of constant (across the gas layers) opacity, and this result

---

[12]Schwarzschild worked on stellar convection during his tenure as professor at the University of Göttingen, in Germany. He is probably most famous for finding in 1915 the first exact solution to Einstein's equations of the gravitational field around a non-rotating spherical massive object, shortly after the publication of the general theory of relativity. This solution envisages the existence of black holes, objects whose gravitational field is so strong that not even light can escape their gravitational pull. He made these calculations while serving on the Russian front during World War I – after volunteering to join the army despite his 40 years of age – as a lieutenant in the artillery division. While in Russia he suffered from a rare autoimmune skin disease and died in 1916. His second son, Martin, became professor of Astrophysics at Princeton university and has played an important role in the development of the theory of stellar evolution, as we will see shortly.

heavily influenced his contemporaries[13]. Eddington acknowledged this result and calculated stellar models considering radiative transport and no convection. He still employed Emden's formalism, but because of the inclusion of radiative transport he also needed to specify how the luminosity changes with the distance from the centre of the model – hence he needed to make assumptions about where the energy was produced within the star, by whatever mechanism – the value of the opacity at each distance from the centre, and add the contribution of the photons to the pressure, while still assuming that the gas particles behave like a perfect gas[14]. He also assumed, as customary at the time, that the chemical composition is uniform within the model.

In the computations Eddington made some daring and with hindsight incorrect approximations – heavily criticised at the time by famous astrophysicists like James Jeans and Edward Milne – about the opacity, and assumed a constant rate of energy generation per unit mass throughout the model. His radiative models predicted the existence of a relationship – not found in fully convective models – between the mass and the luminosity of stars, independent of the nature of the process that generates the stellar energy, the same for both dwarfs and giants. The luminosity of a star was predicted to increase with its mass, and based on the observations available at the time[15], Eddington found that his predictions matched the data qualitatively, but with an offset that pointed to a systematic error in the opacities or in the choice of the chemical composition of the models.

Irrespective of this offset, Eddington's result went against the evolutionary theory of Russell, for the following reason. According to Russell, at the transition between dwarfs and giants, the matter in stars stopped behaving like a perfect gas. However, data for stars with masses both in the giant and dwarf regions conformed to Eddington results, that assumed perfect gas for all models. If the gas in the dwarfs is behaving differently, some discontinuity or sharp change of the slope of the mass-luminosity relation should appear in the data, but this was not the case.

---

[13]He suggested that if you went farther down into the Sun, you might find convection.

[14]The eminent astrophysicist James Jeans didn't think that the dense cores of stars could be approximated as a perfect gas. He believed the atoms were so closely packed that they behaved like a liquid, rather than a gas.

[15]The data available to Eddington covered a range of masses between about 0.2 and 25 times the mass of the Sun.

Eddington saw two main possibilities to move forward. The first one was more conservative, and he embraced it at first. Russell's HRD was considered so far as an evolutionary path: Stars are born in the region of the giants and evolve to the top end of the dwarf sequence first, before cooling down along the observed diagonal path in the diagram, towards fainter luminosities.

The idea that the HRD of stellar populations represented an evolutionary path of stars was so ingrained that also the HRDs of star clusters were being interpreted in this framework, even though they look quite different from Russel's HRD. Figure 3.4 shows the first-ever HRD by Rosenberg, already mentioned before. It shows stars in the Pleiades cluster and its morphology is clearly different from Fig. 3.3. In his pioneering work on Milky Way globular clusters published between 1916 and 1920, Harlow Shapley noticed that the brightest stars (the stars he could observe with the telescopes of the time, given the larger distances of globular clusters compared to open clusters) were mainly on the cool side of the HRD. No bright stars were hot, contrary to Russell's HRD. In 1925 Robert Trumpler analysed the HRDs of 52 of open clusters, and noticed that, compared to Russel's HRD shown in Fig. 3.3, the open clusters emphasized some sequences more strongly, while others were missing. To explain these differences, Trumpler assumed – correctly, as we will see in chapter 7 – that all stars in a cluster form at the same time, and that different morphologies likely correspond to different ages. However his scenario to explain the HRDs in terms of an evolutionary sequence following Russell's ideas led him to conclude – erroneously, as we know today – that more massive stars evolve on longer timescales, and that some clusters comprised more massive stars than others, at the same age.

To 'save' Russell's picture of the HRD as an evolutionary path, Eddington considered the following scenario. The giant evolution was as envisaged by Russell, however when stars reach the dwarf sequence a hypothetical[16] electron-proton annihilation named *mutual suicide* by Eddington sets in, whereby the mass of the two particles – essentially the mass of the proton, about 2000 times larger than that of the electron – is transformed into energy[17].

---

[16] This is incorrect, as we now know, because the proton is not the antiparticle of the electron, and annihilation happens only between a particle and its antiparticle, see the appendix.

[17] Jeans claimed he had introduced the idea of annihilation in astrophysics before Eddington, although he opposed Eddington's proposal of this process being the source of the stellar energy.

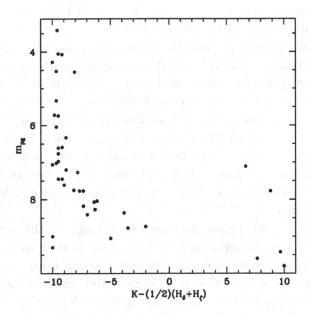

Figure 3.4 HRD of the Pleiades using the same data Rosenberg published in 1910. The horizontal axis displays a quantity similar to spectral type: Cooler stars have larger values of the horizontal coordinate. The magnitude (a proxy for the luminosity) on the vertical axis is not corrected for the cluster distance.

As a consequence the mass of the star, hence its luminosity according to Eddington mass-luminosity relation, is reduced with time during the evolution along the dwarf sequence [18]. No explanation for the transition to the proton-electron annihilating dwarfs was given. This idea of *mutual suicide* of a proton and an electron was considered seriously throughout the 1920s, at a time when the concept of antimatter wasn't developed yet. Even two giant figures in theoretical physics like Niels Bohr and Paul Dirac (who introduced the idea of antimatter) discussed the annihilation of an electron and proton pair in an exchange of letters in 1929.

If the mass of the annihilated protons is transformed into energy according to Einstein formula $E = mc^2$ (where m is the mass of the proton and c the speed of light), the amount of energy available to be

---

[18]In his *The Internal Constitution of the Stars* Eddington denoted as Main Series the dwarf sequence in the HRD.

radiated away by stars is such that they would shine for trillions (a trillion is a number equal to 10^{12}) of years.

The second possibility was that Russell's does not represent an evolutionary track, rather a locus of equilibrium points, a snapshot of where stars with different mass are located at this point in time in the HRD. This also implies that stars with different mass do not necessarily evolve along the same path in this diagram. Within this framework, stars remain of nearly constant mass throughout their lifetime. As for the energy source, the idea of element building, especially transformation of hydrogen to helium – the next element in order of increasing number of particles in its nucleus – was starting to be considered as a serious possibility.

As early as 1899, before any real knowledge of the structure of atoms, Thomas Chamberlin presciently wrote that despite our ignorance, it was not improbable that atoms are the seats of enormous energies, possibly comparable to energies that reside in gravitation. Just 21 years later Francis Aston was able to determine experimentally that the mass of helium nuclei was slightly less than the mass of four hydrogen nuclei (four protons). Thus, if four protons could be somehow converted to helium, the mass difference could be released as energy again according to $E = mc^2$. At the time, only protons and electrons were known; the neutron, the other constituent of atomic nuclei with a mass almost identical to that of the proton (a helium nucleus is made of two protons and two neutrons instead of four protons), had not been discovered yet. However, Eddington models were calculated for a gas made of iron, like it was assumed for the Earth core at the time, hydrogen wasn't considered abundant enough to be included in model calculations.

This composition was soon to be changed to solve the problem of the offset between the theoretical and observed mass-luminosity relation, thanks also to the indications from spectroscopy. Scientists were starting to understand how absorption lines form after Bohr's theory of atomic structure and work by Meghnad Saha, Ralph Fowler, and Edward Milne on the ionization state of atoms (how many electrons are around the nuclei of the various atomic species, when varying the temperature and density of the gas). They could now attempt to quantify the abundances of the various elements whose lines appeared in the spectra of stars.

In her landmark 1925 PhD thesis, considered by Otto Struve – one of the most prolific astronomers of the twentieth century – the most brilliant Ph.D. thesis yet written in astronomy, Cecilia Payne-Gaposchkin showed how to determine effective temperatures from the strength of spectral lines, and even more importantly, she made a revolutionary discovery regarding the chemical composition of the Sun. Elements in the solar spectrum were generally present in about the same relative amounts as on Earth, in agreement with the common view of the time that stars had approximately the same elemental composition as our planet. But helium and particularly hydrogen were much more abundant: This was so unexpected that Russell, in a letter to Payne-Gaposchkin dated 14 January 1925, wrote that he was convinced there was something seriously wrong with the present theory of line formation. He continued writing that it is impossible hydrogen is a million times more abundant than the metals, and must have been very persuasive[19], for she wrote in her thesis that the results for hydrogen are probably the manifestation of abnormal behaviour. Eventually, Russell was persuaded in 1928 of the reality of the large abundance of hydrogen, after similar abundances were also reported by Albrecht Unsöld in the solar atmosphere, although Unsöld himself didn't believe that these abundances would apply to the Sun interior. Because hydrogen is so light (nucleus made of just one proton), he thought it would float to the surface of a star.

Seizing upon these indications from spectroscopy, in 1932 Eddington modified the composition in his calculations by assuming that about 30-35% of the mass of his model stars was made of hydrogen, solving the discrepancy of his mass-luminosity relation with observations. The same year Bengt Strömgren found similar results, by calculating stellar models with varying compositions to match the observed radius and luminosity of a sample of stars with known mass.

Around the same time, the discovery of the cosmic expansion by Lemaitre and Hubble suggested a way to determine the age of the universe. Assuming as a first approximation that the Hubble constant H_0 has stayed the same since the Big-Bang, the relationship $v = H_0 D$ implies that the ratio between any galaxy distance D and its velocity of recession v (equal to $1/H_0$) corresponds to the time elapsed since the moment all galaxies were squeezed together at the Big-Bang. The value of H_0 determined by Hubble in 1929 gave an age of the universe on the order of 2 billion years, incompatible with the idea of electron-

---

[19]She told this to Owen Gingerich, as reported in his paper listed at the end of this chapter.

proton annihilation. Thus, the only option for Eddington to explain Russell's HRD and the mass-luminosity relation was to assume that the observed sequences are a locus of equilibrium points, and that nuclear transmutation powered the stars. The stage was now set to finally identify the source of the stellar energy.

3.3   THE ENERGY SOURCE

As early as 1919, Russell set out some prerequisites for the *unknown process* (his words) that powers stars, as required to solve the discrepancy between the age of the Sun (determined from gravitational contraction) and the much longer timescales needed by geology [101]. Two of these requirements are particularly important. First of all, the energy generated by the *unknown process* must be released near the centre of the star, to keep the core hot and generate a pressure large enough to keep the star in hydrostatic equilibrium. The second requirement is that the energy supply must be strongly dependent on temperature, to keep the star stable. If the star is compressed out of equilibrium by its own gravity (hence the gas heats up), energy production needs to be boosted to increase the pressure forces and make the star spring back to the original equilibrium radius. This happens only if the temperature increase due to compression raises the energy production fast enough to prevent the total collapse of the star. Conversely, if the star expands out of equilibrium (hence the gas cools down) because too-high gas pressure, the energy supply must decrease quickly enough to stop and reverse the expansion.

In his Presidential Address to the 1920 annual meeting of the Mathematical and Physics Section of the British Association for the Advancement of Science, Eddington referred to Aston's measurements of the difference between the mass of four protons and of one helium nucleus, and stated that if just 5 per cent of the mass of a star consists initially of hydrogen atoms, which are gradually being combined to form heavier elements, the search for the source of the stellar energy was over.

The problem was that, at the temperatures calculated for the cores of stellar interiors, the electrostatic repulsion between protons (electric charges with the same sign repel each other) was too high to be overcome by their kinetic energy. How was it then possible to bring hydrogen nuclei close enough for nuclear reactions to become effective? The key to the solution came in 1928 with the work of George Gamow, who applied the recently discovered laws of quantum mechanics to calculate the probability of interaction between particles with the same electric charge. Gamow's result was applied to astrophysics by Robert Atkinson and Friedrich Houtermans, who in 1929 showed that at the temperatures of the cores predicted by Eddington models (on the order of 10 million Kelvin) a small fraction of particles can overcome the Coulomb electrostatic repulsion, and produce heavier elements (mainly helium) and energy, even though the precise chain of reactions wasn't yet known. Atkinson also found in 1931 that the nuclear energy generation in stars must have a very strong dependence on the temperature T, as required by Russell.

In 1932 James Chadwick discovered the neutron, improving our knowledge of the nuclei of atoms, and physicists started to find a large number of proton-induced nuclear reactions from their experiments. On the side of stellar modelling, between 1932 and 1936 Ludwig Biermann and Thomas Cowling independently applied the criterion derived by Schwarzschild in 1906 to assess whether convection is present in a gas layer within a star, and determined that the cores of stars are convective if the energy generation is concentrated at the centre (meaning when the energy generation varies as a high power of T), and that an appreciable part of the envelope can also be convective in some type of stars. Thus, most likely real stars were neither fully convective as Ritter and Lane models, nor fully radiative, as Eddington models.

By 1939, thanks to the works by Carl von Weiszäcker and especially Hans Bethe, the chain of nuclear reactions leading to the fusion of four hydrogen nuclei to produce a helium nucleus was discovered. Two different sequences of reactions are efficient in producing helium, depending on temperature, the 'proton-proton chain' (p-p chain) and the 'CNO-cycle', this latter involving proton captures on pre-existing carbon, nitrogen and oxygen nuclei. Calculations of models for stars on the dwarf sequence in the HRD – by this time commonly named main sequence, because it contains most of the observed stars – predict helium production in the core through the CNO-cycle above a certain mass, that today is calculated to be around 1.0–1.2 solar masses (the exact value depends on the initial chemical composition of the star). Less massive models are powered by the p-p chain. The CNO-cycle energy generation has a stronger dependence on the temperature, the energy production is very concentrated at the centre of the models, and cores are convective, whilst they are radiative in case of the p-p chain.

3.4   GIANT STARS AND WHITE DWARFS

The discovery of the energy generation mechanism in main sequence stars was a triumph of stellar astrophysics. Finally, the lifetime of the Sun turned out to be on the order of billions of years, solving the conflict with the age of the Earth required by geology and biology.

Another major success of those years involved two faint stars in Russell's HRD of Fig. 3.3. If we look carefully, we will notice two objects of spectral type A – hence with a surface temperature hotter than the Sun – about 9–10 magnitudes fainter than the average magnitude of main sequence stars of the same type, and about 5 magnitudes fainter than the Sun. These two stars are 40 Eridani B and Sirius B, both binary companions of brighter objects.

This magnitude difference means that their luminosities are about 10,000 times fainter than main sequence stars with approximately the same surface temperature. Given that at a fixed surface temperature the luminosity of a star scales as the square of its radius, these two objects have a radius roughly 100 times smaller than their main sequence siblings with the same spectral type. And if their mass is approximately the same as their main sequence counterparts, the corresponding mean density (that scales as the inverse of the radius cubed) is roughly a million times larger!

At the time, the analysis of the Sirius binary system provided a mass of 0.85 solar masses for Sirius B (the current measurements give a value equal to about the mass of the Sun), as obtained through the measurement of the mass ratio to the main sequence companion Sirius A, together with an estimate of Sirius A mass. The radius of Sirius B (obtained again from the study of the binary system) was estimated to be 21,000 Km (current estimates give about 5,800 Km, roughly 90% of Earth radius), resulting in a mean density about 40,000 times the mean density of the Sun (the modern estimates of mass and radius result in a density about 50 times higher than this value). This means that a teaspoon of Sirius B material here on Earth would typically weigh about 250 Kg!

The existence of these faint objects with small radii and extremely high densities, firstly named white dwarfs by Willem Luyten in 1922, was puzzling. How could they stay in hydrostatic equilibrium at such small radii?

In 1926 Ralph Fowler considered some recent results of quantum mechanics within the context of stellar interiors. Inside stars atoms are generally ionized due to the high temperatures, meaning that their electrons have been stripped from the nuclei, and the gas is made of bare atomic nuclei plus free electrons. When the density is extremely high, quantum mechanics predicts for the sea of free electrons a very odd phenomenon, named degeneracy. Basically, when degeneracy sets in, no two identical electrons may share the same energy. As a consequence, the electron-degenerate gas is more resistant to compression than a perfect gas, and its pressure is independent of temperature, increasing with the density ρ as $\rho^{5/3}$. Fowler argued that this pressure would be enough to maintain the white dwarfs in hydrostatic equilibrium despite their very compressed state, hence the very strong gravitational force acting upon the gas.

Between 1930 and 1935 the works by Edmund Stoner and Subrahmanyan Chandrasekhar confirmed that white dwarfs are indeed supported by the pressure of degenerate electrons, and obey a well-defined relationship between their mass and radius, more massive objects having smaller radii (hence they are denser). These authors, working independently, also realized that when densities are sufficiently high (this implies that the degenerate electrons reach higher energies), the effects of special relativity on the degenerate electron pressure need to be taken into account. It turned out that in this regime, the pressure increases with increasing density as $\rho^{4/3}$, slower than $\rho^{5/3}$, and as a consequence, the mass of white dwarfs has an upper limit, called Chandrasekhar mass, equal to about 1.4 solar masses[20]. At the densities of this limiting mass the electron pressure can just balance gravity, but the increase of pressure with density cannot keep up with the increased gravity when the white dwarf mass rises above this limit, hence hydrostatic equilibrium cannot be maintained. The astrophysical consequences of the existence of this limiting mass weren't understood straight away, nor it was known how nature produced these puzzling objects. Eddington was fiercely opposed to this idea, most likely because he would not accept that a star could collapse indefinitely, without having a chance to achieve hydrostatic equilibrium. After Chandrasekhar presented his results at a Royal Astronomical Society meeting in London in January

---

[20]Chandrasekhar's original paper gave a value of 0.91 solar masses, because he considered an incorrect chemical composition for a typical white dwarf. Stoner also determined the existence of this limiting mass, with a slightly different value because of his incorrect assumption of constant density inside white dwarfs. Stoner discovery predates Chandrasekhar result, but his work has been somewhat forgotten, maybe because he then left astrophysics to focus on theoretical physics.

1935, Eddington claimed that there should be a law of nature to prevent a star from behaving in such an absurd way. Regarding the inclusion of the effects of special relativity in the calculation of the degenerate electron pressure, Eddington thought that it was a combination of special relativity and quantum theory that he did not regard as *born in lawful wedlock*. In 1983, Chandrasekhar was eventually awarded the Nobel Prize for *his theoretical studies of the physical processes of importance to the structure and evolution of the stars*, among which the work on white dwarfs played a major role.

While reviewing these recent advances in a 1939 article published in the *Journal of the Royal Astronomical Society of Canada*, Russell speculated that when hydrogen is exhausted, a star contracts, becoming fainter and smaller, and end up, if of small mass (like the Sun for example) as a white dwarf.

This was the proposed (but incorrect, as we now know) evolution of stars like the Sun, but what about the giants? It is important to realize that with the Lane-Emden equation, the internal composition had to be changed 'by hand' from one model to the next, and the structure recalculated with the new chemical abundances. The corresponding age change Δt was then determined by calculating how much energy E_H is produced by the amount of hydrogen converted into helium, to compute the ratio $\Delta t = E_H/L$, where L is the luminosity (the amount of energy released per unit time) of the last model.

In this way, Strömgren had shown that the gradual transformation of hydrogen to helium during the main sequence evolution shifts the models towards hotter surface temperatures, if they were kept of homogeneous composition, not towards the cooler giant region. This assumption of uniform chemical composition was still maintained, even for models calculated without convection, and followed from Eddington solution of the problem highlighted by Edvard von Zeipel. von Zeipel showed that radiative regions in a rotating star[21] cannot be in equilibrium, but Eddington (and independently Heinrich Vogt the same year) found that the problem can be solved if 'rotational currents' of gas (large scale gas motions across the star) transport energy and matter across the radiative layers. According to Eddington, these rotational currents would always keep the chemical composition uniform across radiative regions, even in case of slowly rotating stars like the Sun,

---

[21] All stars rotate, with rotational velocities that display large star-to-star variations.

whose rotation period is of about 26 days at the equator, and around 35 days in the polar regions[22].

In 1938 Ernst Öpik[23] wrote two seminal papers titled *Stellar structure, source of energy and evolution* and *Composite stellar models* respectively, published in the almost inaccessible Estonian journal *Publications de l'observatoire astronomique de l'université de Tartu*, where they remained almost unknown to coeval scholars. In these papers, he concluded that convection can be present in stars, although it would not mix them completely. Also, rotational currents are not able to mix efficiently the whole star; as a consequence, given that energy generation increases with increasing temperature, on the main sequence hydrogen is exhausted first near the centre and stars would tend to develop in time a non-uniform chemical composition. He showed that it was possible to calculate models of giant stars assuming a dense core depleted of hydrogen because of the nuclear reactions during the main sequence phase, surrounded by hydrogen-rich layers with hydrogen-burning still going on (the hydrogen-burning shell), and a more external and cooler envelope untouched by nuclear reactions. The existence of an inert (without ongoing nuclear reactions) denser helium core, and a contrast in chemical composition between the heavier helium core and the

---

[22]The Sun is not a solid body like for example Earth, and unlike our planet the rotation period depends on the latitude

[23]The eventful life of Ernst Öpik is a mirror of the turbulent history of Europe during the twentieth century. Öpik was born in Estonia in 1893, and developed an awareness of the beauty of the cosmos, thanks to his older sister Anna. At that time Estonia was part of Tsarist Russia, and after completing his high school studies in Tallin, he went to study at the University of Moscow in 1912. During the Russian revolution, he opposed the Bolsheviks, was captured and sentenced to death, but somehow managed to be released. In 1919 he was sent to work at a new university set up in Tashkent, in Uzbekistan; when Estonia gained its independence from Russia two years later, he went back to his native land, at the Tartu Observatory, where he finally obtained his PhD two years later. Öpik remained in Estonia under the German occupation in the Second World War but left for Germany in 1944 when the Red Army arrived. He lived in a refugee camp in Northern Germany for several years, serving as Rector of a university set up in Hamburg for refugees of former Baltic States. He finally moved to Northern Ireland in 1948 and worked at the Armagh Observatory until the end of his career (he died in 1985), living through the times of the Northern Ireland conflict. During his whole life, even when in a refugee camp, he continued to publish astrophysics papers. I found 430 publications listed to his name, often published in lesser-known journals; they span a broad range of topics, from solar system research to star formation and stellar evolution, galaxies and cosmology.

external layers richer in lighter hydrogen were crucial ingredients to explain the giant stars in Russell's HRD.

Several authors, like Fred Hoyle and Raymond Lyttleton first, followed by Allan Sandage and Martin Schwarzschild[24], rediscovered and pursued this avenue of research in the following years, while Mario Schönberg and Chandrasekhar studied the evolution of the helium cores left over after the main sequence. In two 1955 papers Hoyle and Schwarzschild were finally able to calculate what are considered to be the first complete evolutionary models of giant stars [55, 56].

Hoyle and Schwarzschild first found that with a helium core (electron-degenerate because of its high density) slowly increasing in mass because of the helium production in the surrounding hydrogen-burning shell, plus a radiative envelope above the shell, models for a star with a total mass 1.1 times the solar value were correctly moving towards the region of the HRD populated by giant stars. However, instead of climbing up in luminosity to be consistent with the position of the giants in the HRD, the models were continuing to move towards decreasing surface temperatures at constant luminosity. The solution to this problem was to assume a convective envelope instead of a radiative one[25]: With this structure, the models eventually began to increase in luminosity with increasing helium core mass, as required by the observations. As an aside, I remember that many years ago I calculated with a modern stellar evolution code the evolution of a model with the mass of the Sun, forcing – because the solution of the full set equations of stellar structure and evolution (see the next chapter) required a convective envelope – the envelope to stay radiative during the evolution: After the main sequence the model indeed evolved indefinitely at almost constant luminosity, towards increasingly lower surface temperatures.

Coming back to Hoyle and Schwarzschild calculations, the growth of the luminosity eventually has to come to a halt, to be consistent with the observations (otherwise the models become too bright). The authors then assumed that due to the increase in the core temperature with luminosity found in their calculations – caused by the rising

---

[24] A perfect example of family tradition: Martin Schwarzschild father was Karl Schwarzschild, and Robert Emden was his uncle. He was only 4 years old when his famous dad died, but Karl Schwarzschild influence steered him towards astronomy at an early age, instead of his other career choice, which was to be a milkman.

[25] At the time also the mechanism of energy transport within the models had to be somewhat input by hand.

temperature of the hydrogen-burning shell – helium-burning (the fusion of helium nuclei to produce heavier elements) is eventually ignited in the centre, at an estimated temperature of 120 million K[26]. This ignition stops the increase of luminosity and removes the electron degeneracy in the core.

These results have been confirmed by modern calculations, and mark the end of this heroic period of stellar evolution studies, when models had to be calculated one at a time by using hand-operated desk calculators, making informed assumptions about the physical conditions of their interiors. It was basically impossible to perform calculations of more complex advanced evolutionary phases in this way, and indeed Hoyle and Schwarzschild themselves wrote in the conclusion section of their second paper, that calculations were becoming so complicated it was now necessary to turn to the use of electronic computing machines.

The same year, Louis Henyey together with Robert Lelevier and Richard Levée calculated the first models of stars using an electronic digital computer (the *UNIVersal Automatic Computer*, or UNIVAC), and in 1959 he presented the numerical method still used today to calculate the evolution of stellar models [49, 48] [27]. Henyey's technique, coupled to the speed and power of the computer to make calculations quickly and accurately, made possible to calculate full self-consistent evolutionary sequences rapidly.

The other major milestone that, coupled to the advent of digital computers, has enabled scientists to investigate and understand later stages of stellar evolution, is the monumental (108 pages) Margaret Burbidge, , Geoffrey Burbidge, William A. Fowler[28], and Hoyle paper published in 1957, titled *Synthesis of the Elements in Stars* [17]. This paper presented the sets of nuclear reactions that can be efficient in stars, to produce all chemical elements we see in nature, starting from hydrogen. Before this titanic work, as early as 1946, Hoyle attempted to explain the origin of the elements other than hydrogen by

---

[26]Helium nuclei contain twice the number of protons of hydrogen nuclei, hence higher temperatures are needed to overcome the stronger electrostatic repulsion.

[27]In 1956 Hoyle and the mathematician Colin Haselgrove performed the first stellar model calculations with the Electronic Delay Storage Automatic Calculator (EDSAC 1) computer in the Mathematical Laboratory at Cambridge, United Kingdom [45]. They used a different technique to solve the equations of stellar structure and evolution.

[28]This work was part of Fowler research on the synthesis of elements, that eventually led to his 1983 Nobel Prize, shared with Chandrasekhar.

nucleosynthesis in massive stars, devising an early – incorrect, as we now know – model for the origin of supernovae[29]. In his theory, after core hydrogen-burning massive stars begin to collapse because gravity can no longer be balanced by the gas pressure. During this collapse, the central regions reach temperatures high enough to produce elements heavier than H through nuclear reactions. These elements are mixed throughout the star, whilst an instability due to rotation ejects the gas (chemically enriched with these heavy elements) from the surface, producing a supernova [30]. Due the increased densities caused by the collapse, the core becomes electron-degenerate and when the mass has been reduced below the Chandrasekhar limit, both collapse and mass ejection stop and the white dwarf remnant evolves in hydrostatic equilibrium.

By the end of the 1950s everything was in place to complete our picture of the internal structure of stars and its evolution in time.

GLOSSARY

Annihilation: Reaction in which a particle and its antiparticle, both with the same mass m, collide and disappear, releasing an amount of energy $E = 2mc^2$. An example is the annihilation between an electron and its antiparticle, a positron.

Blackbody: Idealized physical body that can be imagined as a cavity in which the radiation is in thermal equilibrium with the walls, meaning that, on average, the walls emit as much radiation as they absorb.

Chandrasekhar mass: Maximum value of the mass for a white dwarf to be able to attain hydrostatic equilibrium.

Convection: Heat transfer due to large-scale motions of fluid particles.

---

[29]Supernovae are blindingly bright stars that appear all of a sudden in the night sky, and then fade away slowly, leaving behind clouds of gas, and sometimes invisible compact remnants. They can briefly outshine entire galaxies and release more energy than stars like the Sun will in their lifetime. He considered an example with a $15M_\odot$ star, a mass equal to the estimated mass of the gas left over by the supernova that appeared in 1054, recorded by Chinese astronomers.

[30]The smaller the radius the faster the stellar surface rotates, like a figure skater spinning, who will bring the arms closer to the body to spin faster. Eventually, the centrifugal force at the surface overcomes gravity and matter ejection ensues.

Electron degeneracy: Consequence of the Pauli exclusion principle in quantum mechanics. In the regime of degeneracy, the electrons are forced to travel at increasingly higher speeds when density increases. The resulting electron pressure depends only on density and chemical composition, and not on temperature.

Hertzsprung-Russell diagram: Diagram showing on the horizontal axis the spectral type (or the surface temperature) of a population of stars, and on the vertical axis their absolute magnitude (or luminosity).

Hydrostatic equilibrium: A state of balance between the force of gravity and gas (or more in general fluid) pressure forces.

Perfect gas: A gas in which all collisions between particles are perfectly elastic, and there are no interparticle forces. The pressure depends on temperature, density, and chemical composition of the gas.

Photosphere: The surface of a star, from where photons with a black-body spectrum are released.

Radiative transport: Heat transfer due to photons.

Spectrum: A recording of the brightness of a light source as a function of wavelength.

FURTHER READING

T. Arny. The Star Makers: A History of the Theories of Stellar Structure and Evolution. *Vistas in Astronomy*, 33:211, 1990

D. Cenadelli. Solving the Giant Stars Problem: Theories of Stellar Evolution from The 1930s to The 1950s. *Archive for History of Exact Sciences*, 64:203, 2009

O. Gingerich. Report on the Progress in Stellar Evolution to 1950. *Stellar populations, International Astronomical Union Symposium no. 164*, Kluwer Academic Publishers, Dordrecht, 3, 1995

P.-M Robitaille. A Thermodynamic History of the Solar

Constitution — I: The Journey to a Gaseous Sun. *Progress in Physics*, 7(3):3, 2011

P.-M. Robitaille. A Thermodynamic History of the Solar Constitution — II: The Theory of a Gaseous Sun and Jeans' Failed Liquid Alternative. *Progress in Physics*, 7(3):41, 2011

G. Shaviv. *The Life of Stars: The Controversial Inception and Emergence of the Theory of Stellar Structure*. Springer, 2009

CHAPTER 4

Introducing the Machine

The development of computers has enabled us to calculate stellar models with much greater speed, detail, and accuracy than before. This, in turn, has facilitated our understanding of the more advanced and complex phases of stellar evolution, to the extent that we now have a fairly complete view of the various evolutionary pathways of stars. As we will see shortly, the picture is more complex than the early ideas of Russell at the beginning of the twentieth century.

Before presenting a concise overview of the current theory of stellar evolution, it is necessary to give a precise definition of what we mean when we say 'stellar models and their evolution', and also provide a general idea of how stellar models are computed today.

We call 'stellar model' a tabulation of the values of physical quantities like for example pressure, temperature, density, but also abundances of chemical elements, or the rate of energy production (how much energy is produced per unit time), from the centre to the surface of a fictitious star with a given mass and initial chemical composition. The 'evolution' of a stellar model means the calculation of how the values of these quantities change with time. These computations require first to translate into mathematics the description of the physics that shapes the structure of a star.

4.1 THE EQUATIONS OF STELLAR STRUCTURE AND EVOLUTION

We are all familiar with equations, more specifically algebraic equations, even if maybe most of the times we are not aware that we are actually using and solving them. In general, equations can be viewed as questions written in mathematical form, so that the rules of mathematics can be employed to answer these questions. In the case of algebraic equations, the answer to our question will be a number for which we know the relationship to other numbers. Let's go straight to an example: I have gone today to the supermarket, but I lost the receipt and do not remember how much I paid. What I do know is that I have now in my wallet a total of 180 British pounds, while in the morning I checked that I had 250 pounds. This information allows me to find out that I spent 70 pounds at the supermarket (I did not pay for anything else during the day), by writing the following simple equation: $250 - x = 180$, where x is the unknown amount of money I spent. My question has been translated into an equation, that I can solve (find the value of x, the answer to my question) using mathematics. Solving an algebraic equation means to use the rules of algebra and calculate the value of the unknown number x, such that the numerical value in the right-hand side of the equation (the expression to the right of the identity symbol '=') is equal to the value in the left-hand side. Of course, algebraic equations can be much more complicated than this simple example, but the concept is the same.

Instead of just one algebraic equation, sometimes we need to solve jointly several of them to answer our question, like in the following example that requires to find out the value of not just one, but two numbers. Let's suppose that I went yesterday to the post office to send 10 parcels weighing 1 Kg each, to five friends in the United States and 5 family members in Italy. Back at home, I got the feeling to have sent one parcel to the wrong destination, but the receipt listed only the total amount I paid, that was 46 pounds, plus the rates, equal to 7 pounds per Kg for deliveries from the UK to the US, and 3 pounds for deliveries to Italy. These three pieces of information together with my question can be translated into a simple system of two algebraic equations, where I named the two unknown numbers x_1 – the number of parcels sent to the US – and x_2 – the number of parcels sent to Italy:

$$7x_1 + 3x_2 = 46$$

$$x_1 + x_2 = 10$$

The first equation states that the total price is equal to 7 pounds times the number of parcels (each parcel weighs 1 Kg) sent to the US, plus 3 pounds times the number of parcels sent to Italy. The second equation states that the total number of parcels sent out is equal to the number of parcels sent to the US plus those sent to Italy. The solution of this simple system of equations, following the rules of algebra, tells me that I sent 4 parcels to the US and 6 to Italy: I definitely sent one to the wrong destination. Notice that I need to have as many equations as the number of unknown quantities, to be able to find a solution to my problem (in this example I could write down two equations for two unknowns).

In stellar evolution the situation is much more complex. We want to determine how physical and chemical quantities change from place to place inside a star of a given mass at a certain point in time, and how they vary over time. Here the unknowns are not numbers, they are functions instead, meaning a relationship between the position within the star and the value of, for example, pressure, or temperature, or the abundance of a chemical element. When the unknown is a function, the class of equations that translate our question into mathematics is called 'differential equations'. Differential equations are routinely employed in natural sciences and also, for example, in finance. Calculations of stellar models require to write down and solve a system of differential equations because the values of several physical and chemical quantities across the whole model need to be determined. And to find a solution we need to specify at the start the total mass and the initial chemical composition of the star we want to model.

The abundances of chemical elements that enter the equations of stellar structure and evolution are typically given as fractions of the total mass of gas. The mass fraction of hydrogen is usually denoted with X, the mass fraction of helium with Y and that of all other heavier elements (named collectively as the 'metals') with Z. By definition, the sum of these mass fractions must be equal to one, hence $X+Y+Z=1$. In case of the initial chemical composition of the Sun (born, as we will see in chapter 7, about 9 billion years after the Big-Bang) the approximate values are $X=0.71$, $Y=0.27$, and $Z=0.02$, whilst right after the cosmological nucleosynthesis $X=0.75$, $Y=0.25$, and Z was negligible.

Let's now summarize the relevant facts we know about stars. First of all, they have a spherical shape, like the Sun; hence all their properties are going to depend only on the distance r from the centre. This is essentially because the strength of gravity doesn't depend on the orientation, just on the distance from the source of the gravitational field.

Given a spherical surface at a distance r from the star centre (like an eggshell, with the shape of a perfect sphere), the properties of the gas are the same across the whole surface. This simplifies enormously the equations because we need to consider only one dimension (parametrized by r), instead of three. Fairly rapidly rotating stars tend to be elliptical (the distance from the centre to the poles is shorter than the distance to the equator), because of the effect of centripetal forces[1], but even in this case the spherical approximation is not too bad, and factors that account for deviations from the spherical shape in case of moderately fast rotation can be 'added' to the same equations of non-rotating models (see below).

We also know that stars are in hydrostatic equilibrium otherwise we would see them shrink in a matter of minutes. To maintain the gas pressure high enough to balance gravity, the energy released in the surrounding space must be replenished by either nuclear reactions or gravitational contraction. This energy then needs to be transported from where it is generated to the surface, where it leaks out of the star.

Finally, let's recall the origin of the spectral lines discussed in the previous chapter. According to Kirchhoff laws, the absorption lines are originated when a blackbody spectrum produced by a hot high-density gas crosses a low-density gas at a lower temperature. A stellar model deals with the hot and high-density gas that emits the blackbody spectrum. A blackbody spectrum, in turn, means that inside the star the gas particles and photons interact so often that they all have the same temperature (see chapter 2).

The surrounding low-density layers where the lines form are the stellar atmosphere; they contain a negligible amount of mass compared to the total mass of the star, and are modelled using a different set of equations. The main reason for this difference is that the flow of photons through the gas needs to be described in different ways when the density is high or low. Here 'high' means that a photon has a 100% probability to interact one or more times with the gas particles, while 'low' means that the probability of one interaction is below 100%. In

---

[1]The centripetal force pushes or pulls an object towards the centre of a circle as it travels, and is required to keep the object in a circular motion. In case for example of the 'hammer throw' in athletics, the hammer is a metal ball attached by a steel wire to a grip. The throwing motion involves four rotations of the body of the athlete, with the ball moving in a circular path: The centripetal force acting on the ball is provided by the tension on the steel wire. In rotating stars, the origin of the centripetal force is gravity, like for the orbital motion of Earth around the Sun.

this latter case the modelling of the flow of photons across the gas layers is much more complicated.

Even in one dimension, the equations that encapsulate this knowledge are complicated and the solution for the values of the pressure or any other quantity as a function of the position within the star, cannot be written as a mathematical formula. Confronted with this situation, we needed a clever way to write the equations and find their solutions. It turned out that what in principle is a system of differential equations can be transformed into a system of algebraic equations, where the unknown quantities to be determined are numbers, not functions. Let's see briefly how this is possible.

Our one-dimensional stellar model can be envisaged as a string, with a beginning, representing the centre of the star, and an end, corresponding to the photosphere. After the total mass M of the model is chosen, this string is partitioned into many small segments[2], enclosed within two consecutive values of the mass coordinate m. For example, the first point along the string corresponds to the centre of the star, hence $m=0$, whilst the last point along the string corresponds to the surface, hence $m=M$, because the whole mass M of the model is enclosed between centre and surface. Intermediate points will be labelled with values of m that correspond to different fractions of the total mass, increasing when moving towards the endpoint of the string. This is shown in Fig. 4.1, where a one-dimensional stellar model is partitioned into a discrete grid of $N-1$ segments. The value of a generic mass coordinate m_i increases from zero when $i=1$ (the centre) to M when $i=N$ (the photosphere). The values of mass, radius, temperature and luminosity at $n=N$ correspond to the total mass M, total radius R, surface temperature (named effective temperature, T_{eff}), and the surface luminosity (the total amount of energy per unit time released in all directions) L. As already mentioned, the radiation released from the photosphere has a blackbody spectrum, and physics tells us that the total energy per unit time and unit area given off by a blackbody with temperature T is proportional to T^4 (the Stefan-Boltzmann law). This leads to the technical definition of T_{eff} of a star (or a stellar model) as the temperature of a blackbody which emits the same amount of energy (summed over all wavelengths) per unit time and unit area. Therefore, there is a relationship between L, R, and T_{eff}, namely $L=4\pi R^2\sigma T_{eff}^4$,

---

[2]In the real three-dimensional world each segment corresponds to a spherical shell.

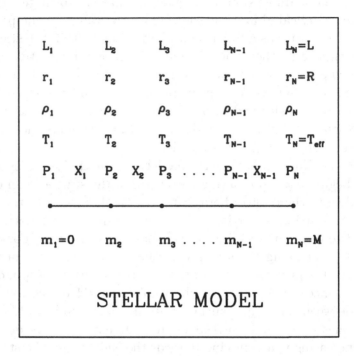

Figure 4.1 Qualitative example of partition of a one-dimensional stellar model into a discrete grid of $N - 1$ segments, between the centre and the surface. At each point i, labelled by the value of the mass m_i, all the various physical and chemical quantities are evaluated. The point with $i = N$ corresponds to the photosphere and mass, radius temperature and luminosity are equal to the total mass, total radius, surface temperature (named effective temperature and denoted with T_{eff}) and surface luminosity (the total amount of energy per unit time released in all directions), respectively. The chemical composition X_i (that here denotes the individual abundances of the various elements) is assumed to be uniform within each segment. The solutions for consecutive segments are connected to each other, because the endpoint of one segment is the starting point of the next one.

where $4\pi R^2$ is the surface area of the photosphere ($\pi=3.14159$ is the ratio of a circle's circumference to its diameter) and σ is a constant (the Stefan-Boltzmann constant, equal to 5.67×10^{-8} when the luminosity is given in Joule per second, the temperature in Kelvin, the radius in metres).

The solution of the equations of stellar structure and evolution for this model of mass M provides values of physical quantities at each point m_i. As for the chemical abundances, they are assumed to be uniform within a segment, and the solution of the equations gives the values of the various abundances for each segment (see Fig. 4.1). Of course, it is important to have a very dense sampling of the string to calculate an accurate model of our fictitious star, and typically we partition the models into hundreds, up to thousands of segments, depending on the evolutionary phase. More advanced evolutionary phases need more segments, because the internal structure of the models is more complex and requires a higher resolution to be accurately described.

It may seem strange that we do not use the radius for our partition – at first glance the most obvious choice – but there is a reason for this. We can think of the stellar gas as a fluid, like water flowing in a river. Calculating physical (and chemical) quantities at discrete values of the distance r from the centre is like to measure the temperature of the water along the river by setting thermometres at specific locations along its path, through which the water flows. On the other hand, using mass to partition the model is like setting the thermometers floating, to follow the various parcels of water as they flow towards the mouth of the river. This second option is particularly advantageous when modelling stars, because we can follow the evolution of the same parcel of gas throughout the whole stellar lifetime, irrespective of any change of the radius .

After having fixed the total mass of the model and determined its partition, we need to specify the initial chemical composition, and a timestep Δt, the time interval between two successive model calculations. The chemical composition is crucial because most of the quantities entering the equations depend on the chemical abundance of the elements that make up the stellar gas. The size of the timestep will depend on the evolutionary phases: Faster phases will need smaller timesteps to be properly modelled.

We are now in the position to write down and solve the equations of stellar structure and evolution for all segments of our model, at a given time t during its evolution. For each segment, the equations to solve are the following:

$$\frac{\Delta r}{\Delta m} = \frac{1}{4\pi r^2 \rho} \tag{4.1}$$

$$\frac{\Delta P}{\Delta m} = -\frac{Gm}{4\pi r^4} \tag{4.2}$$

$$\frac{\Delta L}{\Delta m} = \epsilon_n - \epsilon_\nu - c_P \left(\frac{\Delta T}{\Delta t} - \nabla_{ad} \frac{T}{P} \frac{\Delta P}{\Delta t} \right) \tag{4.3}$$

$$\frac{\Delta T}{\Delta m} = -\frac{T}{P} \nabla \frac{Gm}{4\pi r^4} \tag{4.4}$$

$$\frac{\Delta X_s}{\Delta t} = f(\rho, T, < \sigma v >, X_r) + f_{mix} \tag{4.5}$$

The left-hand sides of the first four equations display the ratio of the difference (Δ), taken between the end and start point of the segment, of various physical quantities – the distance from the centre r, gas pressure P, luminosity L (energy released per unit time), and gas temperature T – to the corresponding difference of the mass coordinate m. These ratios are positive if the values of these quantities increase towards the surface. The right-hand side of the same equations contain quantities calculated as the average of the values at the beginning and the end of the segment.

Equation 4.1 is simply a better way – from the computational point of view – to state that the mass enclosed in a thin spherical shell within the star is proportional to the volume of this shell and the local density of the gas. Equation 4.2, that includes the gravitational constant G, is the condition of hydrostatic equilibrium, meaning that at each point in the model, the gravitational attraction is balanced by the outward pressure of the gas, while equation 4.3 links the production of energy to the luminosity. We usually think of luminosity as the energy given off by the star per unit time, its surface luminosity. Indeed this is what I generally mean throughout this book, but luminosity is more generally defined as the rate of the flow of energy across a segment of our model star. At the outer segment, this corresponds to the amount of energy released by the star per unit time. Therefore, each segment of a stellar model has its own luminosity, calculated through this equation. The term ϵ_n in the right-hand side represents the amount of energy per unit mass and unit time released by nuclear reactions. Its value depends on the abundances of the elements involved in the reactions, temperature, density, and probabilities for the various reactions to happen (quantified by the value of the nuclear cross sections $< \sigma v >$ for each active reaction). The term ϵ_ν represents the amount of energy carried by neutrinos, and depends also on the chemical composition, temperature and density. Neutrinos are produced by either nuclear reactions or other processes related to the weak force (see the appendix). The energy carried by neutrinos is 'lost' by the star, because neutrinos do not interact with the gas particles (hence the minus sign in front of this term). Their energy does not help the gas to keep the pressure high enough to balance gravity, because it is not transferred to the gas particles.

The last terms on the right-hand side of this equation contain the ratios of the differences of the average pressure and temperature within the segment, to the timestep Δt, plus the so-called specific heat c_P – a measure of how much energy is needed to raise the temperature of the gas by one degree – and the symbol ∇_{ad} – the adiabatic temperature gradient – a quantity that can be determined from the values of the gas pressure, temperature and chemical composition. These terms calculate the amount of energy absorbed or released by the gas when the segment is compressed, or expanded and/or its temperature changes.

The following Eq. 4.4 details the effect of the transport of energy on the temperature distribution. The symbol ∇ denotes how much the temperature changes for a small pressure change within the segment (temperature gradient). Its mathematical formulation depends on how the energy is carried across the star. There are three possible modes of energy transport in stars, two of them already mentioned in the previous chapter: Convection, radiation and conduction. Convection means that energy is carried across the different layers by the motion of the gas itself. Gas bubbles from hotter layers rise and eventually dissolve, releasing their excess energy in the environment, while cooler gas is displaced and flows downwards, gets hotter and then raises, in a continuous cycle.

Radiative energy transport is mediated by the photons coming from the energy generation regions. It is always efficient as long as a temperature difference does exist. In their travel towards the surface, these high-energy photons suffer innumerable collisions with the gas particles, an endless sequence of scattering, absorption and reemission, which randomly alter their direction and redistribute their energy to the gas particles. Their path is like a random walk, and the progress towards the surface is quantified by the value of the 'opacity' of the gas (usually denoted with the symbol κ), the probability that the photons experience one interaction with the gas particles in a unit length: The higher the opacity, the slower the progress because of more interactions with the gas. If the motion of the photons towards the surface is too slow compared to the amount of energy to redistribute across the star, convection sets in. This is the so-called Schwarzschild criterion, that tells us when convection is the dominant energy transport in a given segment of our stellar model.

The last energy transport mechanism is conduction. Conduction is the transfer of heat through the star by the collisions among gas particles. More energetic particles travel across the star and redistribute their energy by colliding with the surrounding, less energetic particles. In a star, this mechanism is efficient only when electrons are degenerate (see the previous chapter), like in white dwarfs or in the interior of red giant branch stars, as we will see shortly. Degenerate electrons can be very energetic compared to the ionized nuclei of the gas, even at relatively low temperatures, and can travel long distances before interacting with the rest of the gas particles. In this way, they are able to redistribute their excess energy very efficiently across the star. The mathematical expression of the temperature gradient is the same as for the radiative transport.

The final Eq. 4.5 comprises actually several equations, one for each chemical element s included in the model calculations. It states that the evolution with time of the abundance X_s (expressed as fraction of the total mass of the gas)[3] of a chemical element s depends on the efficiency of the nuclear reactions that produce or destroy this element, plus other processes that can alter the chemical composition of the star. The left-hand side contains the ratio of the variation of X_s to the timestep Δt, while the right-hand side – not written down explicitly, because it has a very complex mathematical expression – describes the processes that can cause this variation. The first term depends on the efficiency of the nuclear reactions that involve element s, which can either produce or destroy this element (and produce energy in the process). It includes

---

[3]Only in this section X_s denotes the abundance of a generic element s, not to be confused with X, that in the rest of the book denotes – as customary – the mass fraction of hydrogen.

the nuclear cross sections $< \sigma v >$, the abundances X_r of the other elements involved in these reactions, gas density, and temperature.

I still remember vividly the first question I was asked at the exam of the stellar evolution course, on a rainy February morning. It was an oral examination, and I was sitting at this small desk, faced by the three examiners, all world-experts in stellar evolution. On the desk I had a few sheets of paper and a pen. Then we started with *Please, write down the equation for the evolution of a generic element due to nuclear reactions*, meaning that I was asked to write the full expression for $f(\rho, T, < \sigma v >, X_r)$. I actually still remember also the last question of that exam, about the Chandrasekhar mass... in between I recall being asked about the progenitors of thermonuclear supernovae (see below), but the rest of the exam questions has been long forgotten.

Coming back to Eq. 4.5, the second term denoted with f_{mix} describes the change of chemical abundances due to other processes, named generically as 'mixing'[4]. The first and most important one is convection, that mixes the gas very fast. The chemical composition is uniform where convection is efficient, because the gas gets fully mixed in a short time (timescales of days, months). Any initial chemical abundance differences are erased, and the final abundances in the whole convective region is an average of the initial values in the segments that are now convective.

Another process included in f_{mix} is 'gravitational settling'. Collisions among the gas particles induce a slow movement of nuclei of different atomic weight, the heavier ones being displaced towards the centre. Essentially all elements but hydrogen tend to slowly move towards the centre, while hydrogen tends to move towards the surface. Opposite to convection, which erases differences of chemical composition and mixes the gas very fast, gravitational settling works towards enhancing chemical abundance differences but over much longer timescales, up to billions of years, like in the Sun. In addition to gravitational settling, also the so-called radiative levitation can be efficient. Radiative levitation means that when photons interact with atoms that retain at least some of their electrons (we say that they are partially ionized) a fraction of the energy of the photons is spent to effectively 'kick' atoms towards the photosphere, counteracting or even reversing the effect of gravitational settling.

---

[4]I have fairly recently written with my colleague Santi Cassisi a very technical review of the physics included in f_{mix} [105].

Finally, whenever stars rotate, chemical elements are also slowly transported from regions where their abundance is initially higher, to regions of lower abundances. The inclusion of rotation in stellar modelling adds more (complicated and somewhat uncertain) terms to f_{mix}, plus an additional equation that describes how the rotational velocity changes with time in each segment of the stellar model. To account for the effect of rotation we also need to add multiplicative factors in Eq. 4.2, and in the mathematical expression for the temperature gradient when energy is transported by radiation. These factors – sometimes called form factors – essentially account for the departures from a perfectly spherical shape when the star rotates, although they are not very significant. Much more important is the chemical mixing when stars spin rapidly and, crucially, how the velocity of rotation changes across the star, because it affects the efficiency of rotational mixings.

All these element transport mechanisms are inefficient inside convective regions, because convection is much faster and erases the effect of any other process, but they can move elements in or out of the boundaries of convective layers.

In summary, the translation into mathematical language of the question we want to address, namely how to determine the physical and chemical structure of a star and its evolution in time, has led us to this system of equations, for each segment of our model. The segments (hence the corresponding equations) are coupled to each other because, starting from the centre, the point that marks the end of each segment corresponds to the initial point of the next one.

To calculate a stellar model we need to solve a system of $4+S$ equations for each segment, where S is the number of chemical elements, and a full model is partitioned into N segments. The unknowns (the quantities we want to determine) are r, P, T, L at each point m_i, plus the S chemical abundances X_s at each segment. We need also to know how ϵ_n (and the nuclear cross sections $< \sigma v >$), ϵ_ν, the opacity for the photons and degenerate electrons, and the various thermodynamical properties of the gas (density, c_P, ∇_{ad}) depend on P, T, and the abundances X_i. As you can see, the number of unknowns is the same as the number of equations, so the problem is well posed and we can search for a solution using appropriate techniques [49, 48, 67]. Today these calculations can be done very efficiently also on laptop computers, and the results can be evaluated quickly with the sophisticated analysis tools available to us. However, the computational efforts during the early computer age were much more heroic[5] as Icko Iben, one of the leading researchers in stellar evolution between the 1960s and 2010, described briefly in his book 'Stellar evolution physics' [60]. He recalls that during the summer of 1960 he began working on calculations of main sequence models using the ILLIAC 1 computer at the University of Illinois. The computer was made of racks of vacuum tubes, and instructions were provided to the machine as punched holes in a paper tape. Any time the computer stopped working a metal key had to be waved in front of a rack of vacuum tubes to restart the calculations. The results weren't numbers in an output file but, again, punched holes in a paper tape, that had to be translated into numbers. Iben recalls that it took several months of computations (lasting several hours a day) to compute just 30 models.

I remember that, when I started my thesis, cupboards and desks around the institute were still filled with stacks of new or used yellow punch cards of the type mentioned by Iben. They were pieces of stiff paper storing digital data represented by the presence or absence of holes in predefined positions[6]. The computer I was employing for my first calculations was of the IBM/370 series, and the huge amount of data necessary for some of the work I was doing – the calculation of tables of opacities to be used in the computation of stellar models – was stored on tape units.

Before closing this section, it is worth presenting briefly a couple of issues that will arise in the rest of this chapter and in the following ones. First of all, observations tell us that stars lose mass from their photospheres, through stellar winds. The Sun loses about $10^{-13} M_\odot$ per year (at this rate, it would take the Sun 10,000 billion years to lose all its mass), a ridiculously small amount, but stars much brighter and hotter, or much brighter and cooler than the Sun, lose mass at substantially higher rates. This mass loss can be accounted for in model calculations by more or less 'simply' modifying the total mass M of the model from one timestep to the next. The problem is that stellar models do not and cannot predict the onset and rates of mass loss, because we cannot follow in three dimensions the dynamics of actual gas particles. Even though we could write down the correct equations, it is impossible to solve them with the current computing capabilities

---

[5]Gravitational settling, radiative levitation, and rotation were not included, because they add additional layers of complexity to the calculations.

[6]Perhaps luckily, I never needed to use them.

and will be impossible for the foreseeable future. The flow of the gas in a star is too chaotic and unpredictable, too 'turbulent' in technical terms, similar to the case of the Earth atmosphere. This is why, for example, long-term weather forecasts are always extremely uncertain.

What we can do is to determine mass-loss rates from observations, or from some simplified model of the external layers of stars, and include them in our calculations, to adjust appropriately the total mass M of the model during its evolution.

Our inability to predict the dynamics of the gas particles across the whole star has consequences also for the term f_{mix} in Eq. 4.5, and the temperature gradient when the energy transport is convection. Here we need to use equations coming from simplified models of convection, suggested by results from laboratory experiments or observations of Earth atmosphere and oceans, where convection plays also an important role (but convection in these settings is not as turbulent as in stars). Chemical mixing due to rotation also needs to be treated with some simplified model to be implemented in stellar evolution calculations.

4.2 A SUMMARY OF STELLAR EVOLUTION

It is sometimes slightly misleading to use terms borrowed from biology, like 'life' and 'death', when discussing the evolution of stars. Even the term 'evolution ' doesn't mean the same as in biology, for it has nothing to do with the changes of inherited traits of a population from generation to generation[7].

We commonly read about stars being born and dying, but it is not clear-cut what 'death ' means for a star. For example, the Sun is going to exist and shine virtually forever, or at least until its temperature (after all nuclear reactions have ceased) will equal that of the surrounding space, the temperature of the CMB (that will be lower in the future) at the time. Let's see then how we can define these terms within the framework of the theory of stellar evolution.

First of all, the equations of stellar evolution can be applied from the moment a star has reached the stage of a hot sphere of gas in

---

[7]However, the biological evolution might be considered to be comparable to the change of initial chemical composition of different generations of stars, caused by the modification of the composition of the interstellar medium due to supernova explosions and stellar winds.

hydrostatic equilibrium. But where do these newly born stars come from?

They are formed out of interstellar matter, that fills the enormous volume of space between stars in galaxies. This interstellar medium (ISM) is made of gas and dust grains, these latter amounting to about 1% of the total mass of the ISM. For example, the Sun and Earth are moving through an ISM containing about 100,000 particles per cubic metre. Within the ISM we can find giant molecular clouds, made mainly of hydrogen molecules (two protons bound together) and dust, with masses typically between 100,000 and one million times the mass of the Sun, and sizes on the order of 10–100 parsec. They are cold, with temperatures of about 10–100 K, and densities up to one million times higher than the ISM density around the Sun.

Stars form in these cold and dense clouds through complex processes than can be sketched, in a very simplified way, as follows. If the density of the cloud is high enough for its temperature, gravity overcomes the pressure due to the random motions of the cloud particles, and the denser parts of the cloud core (the core is denser than the outer cloud) will collapse under their own gravity. As these regions are squeezed, they fragment into smaller clumps of matter. These small clumps eventually stop being squeezed by gravity when their density is high enough for the energy gained by the gravitational contraction to be trapped within the clumps and fight the force of gravity. At this stage, these clumps settle in hydrostatic equilibrium, and we can begin to calculate stellar models using the set of equations discussed before[8]. Star formation in molecular clouds produces clumps, hence stars, of different masses. However, we are not yet able to predict from first principles how many stars form in a cloud with a given total mass, chemical composition, density, and temperature, and their masses. A short synopsis of how stars evolve, as derived from the computation of stellar models, reads like this.

The evolution is driven by the loss of energy from the surface, which works towards breaking the hydrostatic equilibrium, and the consequent readjustments aimed at replenishing the star energy content and rebalance gravity.

To stay in hydrostatic equilibrium a star needs to be hotter and denser in the core compared to the surface, to create a trend of

---

[8]This star formation phase lasts a time that is negligible compared to the evolution that follows.

increasing pressure when moving towards the centre; as a consequence of the temperature gradient, heat (energy) flows from the centre to the surface of the star. Right after formation temperatures are not high enough to start nuclear reactions, and at this stage the star contracts slowly and heats up, according to the virial theorem: This is the so-called pre-main sequence phase (PMS). The internal temperatures eventually reach values on the order of 10 million Kelvin, able to ignite the fusion of hydrogen to helium in the core and produce energy. This marks the start of the core hydrogen-burning phase, or main sequence (MS), when displayed in a HRD (discussed in the next section). It corresponds to the dwarfs' sequence in the HRD of Russell (see the previous chapter).

The energy produced by nuclear reactions is enough to continuously restore hydrostatic equilibrium, despite the loss of energy from the surface. When hydrogen is exhausted in the central regions[9] – whose chemical composition becomes essentially pure helium – the core begins again to contract according to the virial theorem. Its temperature increases, while the fusion of hydrogen into helium shifts to a surrounding, thin (containing a small fraction of the total mass of the star) spherical shell around the helium core.

During this shell hydrogen-burning the mass of the helium core slowly increases due to the fresh helium produced by the surrounding shell (this stage corresponds to the 'subgiant branch' – SGB – and 'red giant branch' – RGB – evolution in the HRD, discussed in the next section). The contraction of the core stops when temperatures in the core reach about 100 million Kelvin, igniting nuclear reactions that transform the helium produced during the MS into carbon and oxygen. This is the core helium-burning phase, lasting until helium is exhausted in the nuclear burning core. At this point the core, now made of carbon and oxygen, contracts again and raises its temperature, until the start of the fusion of carbon into heavier elements (at temperatures around 600 million Kelvin), whilst the fusion of helium shifts to a shell around the core (and an active hydrogen-burning shell surrounds this helium-burning shell). If the mass of the star is higher than about 10 $M_\odot$, this sequence is repeated several times after the carbon in the core is exhausted, leading to the burning of progressively heavier elements (with an increasing number of protons and neutrons in their nucleus) that were essentially the products of the previous burning stage.

---

[9]For the Sun, this will happen in about 6.5–7 billion years.

This sequence ends when the core is made of iron. The interior of the star is now so hot and dense to produce neutrinos very efficiently, causing a substantial loss of energy. Also, electron degeneracy sets in the iron core, whose mass is larger than the Chandrasekhar mass. This causes its collapse (hydrostatic equilibrium cannot be maintained) leading to the ejection of the surrounding layers, which produces a so-called core-collapse supernova. What is left of the iron core is a neutron star of 1.4–2.0 $M_\odot$ (an incredibly dense object with a radius of a few kilometres, where gravity is balanced by the pressure of neutron degeneracy, a phenomenon similar to electron degeneracy) or a black hole if the mass of the iron core was larger than 2.0–2.5 $M_\odot$. Most of the chemical elements synthesized during the previous evolution (for example helium, carbon and oxygen) are ejected into the ISM by the supernova explosion.

If the initial mass of the star is lower than about 10 $M_\odot$ but higher than about 0.5 $M_\odot$, the sequence of core burnings stop after helium is exhausted in the core if the mass is below about $6–7M_\odot$[10], or at the end of core carbon-burning, for higher masses. The reason is that electron degeneracy sets in the resulting C-O (meaning carbon and oxygen) – for masses below $6–7M_\odot$ – or O-Ne (meaning oxygen and neon) – for higher masses – cores, and this prevents their contraction to increase the temperature until the next nuclear burning is ignited (plus a large production of neutrinos subtracts energy from the cores). Observations tell us that stars in this phase (called 'asymptotic giant branch' – AGB) lose mass from the surface through strong stellar winds. Eventually, almost the whole envelope around the electron-degenerate cores is lost, leaving behind a white dwarf, with mass much lower than the initial mass of the star. The Sun will become a white dwarf in about 8 billion years from now, with a mass expected to be around 0.55 $M_\odot$, about half its present value (see chapter 8). Stars with masses around 6–7 $M_\odot$ will become white dwarfs with a mass about the mass of the Sun, and a 9-10 $M_\odot$ MS star will end up as a O-Ne white dwarf with mass around 1.2 $M_\odot$. Mass loss before the white dwarf phase prevents the formation of objects with masses above the Chandrasekhar mass.

Given the much larger range of initial masses that produce C-O degenerate cores compared to O-Ne ones, the overwhelming majority of white dwarfs are indeed C-O white dwarfs. Their evolution is a slow

---

[10]The exact values of these threshold masses depend on the initial chemical composition

and indefinite cooling: The degenerate electrons provide the pressure sufficient to maintain hydrostatic equilibrium (electrons cannot lose energy because of the quantum mechanical 'degeneracy'), whilst the luminosity is provided by the kinetic energy of the ions that is slowly radiated away (more in chapter 8).

If the mass of the star is between about 0.1 and 0.5 $M_\odot$ electron degeneracy sets in already in the helium core at the end of the MS, and the core never reaches temperatures high enough to start the helium-burning[11]. The hydrogen-burning shell increases the mass of the helium core after the MS, until it reaches the cooler surface (essentially the whole star has been converted to helium) and switches off because of the low temperatures. These stars end up as helium white dwarfs, that follow the same evolution of C-O and O-Ne white dwarfs. The MS lifetime of these objects is equal to tens up to hundreds billion years, and no helium core white dwarf has been yet produced this way.

Objects formed with masses between 0.1 $M_\odot$ and a few times 0.001 $M_\odot$ are named brown dwarfs: They become electron-degenerate during the PMS and never reach temperatures high enough to start the MS, eventually cooling down in the same way as white dwarfs. Given that brown dwarfs never start hydrogen burning, they are not considered to be proper stars, even though they generate their own luminosity by releasing the kinetic energy of their ions.

Figure 4.2 shows simplified representative sketches of the chemical structure of models in the different mass ranges discussed before (bar the brown dwarfs), when they are just formed, at the end of the MS, and at the final stage of evolution, respectively. The chemical composition at formation is always homogeneous, mainly hydrogen plus much smaller percentages (in mass fraction) of helium and metals, because stars are fully convective at this stage. At the end of the MS the chemical structure is still essentially the same in all mass ranges, with a core made of helium (plus the small mass fraction of metals) and an envelope made mainly of hydrogen, unaffected by the nuclear reactions. The different evolutionary paths show themselves in the final configuration which is either a He, C-O, or O-Ne white dwarf or, for the upper range of initial masses, a neutron star or black hole. In this latter case, the figure displays the structure of the models right before the

---

[11]Stars between 0.5 and about $2.3 M_\odot$ also develop degenerate helium cores, but calculations show that they manage to ignite helium in the so-called helium flash at the end of the RGB phase, which also removes the electron degeneracy.

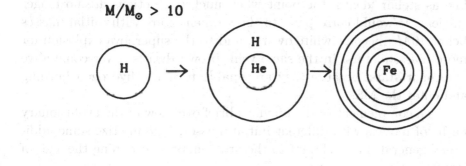

Figure 4.2 Chemical structure of models at selected stages during their evolution, in the three labelled mass ranges. The labels H, He, C, O, Ne, Fe stand for hydrogen, helium, carbon, oxygen, neon, and iron, respectively.

supernova explosion which then leaves a neutron star or a black hole as a remnant. These models have a core made essentially of iron, surrounded by several shells with chemical compositions dominated by the elements involved in the previous burning stages. The more external the shell, the lighter (in terms of atomic weight) the element, corresponding to increasingly early burning stages. The uppermost layers are made mainly of hydrogen, and have never been involved in nuclear reactions. The neutron star or black hole remnants originate from the iron core, whilst the more external concentric shells are expelled during the supernova explosion, and pollute the ISM, from which new generation of stars are going to form. The oxygen that makes up 65% of our bodies is essentially all produced in the interiors of these core-collapse supernovae.

Also lower mass stars that become either C-O or O-Ne white dwarfs shed their external layers before the final white dwarf stage, and the gas they return to the ISM has a chemical composition different from the initial one. Although they have not been directly involved in nuclear burnings, some elements produced by nuclear reactions have been dredged up to the surface – mainly by convection – during the previous evolution. For example, most of the carbon and nitrogen that make up about 20% of our bodies, have been spread into the ISM by stars that are currently cooling down as white dwarfs.

Within this general picture of the evolution of stars[12], we may define as stellar 'death' the point when nuclear reactions are no longer efficient. It would correspond to the moment stars with initial masses below $10M_\odot$ become white dwarfs, and to the supernova explosion for more massive stars. In the same vein, brown dwarfs never come alive as proper stars, because they are unable to ignite hydrogen burning and enter the MS.

As a complement to the previous brief overview of the evolutionary paths of models with different initial masses, I summarize some additional general rules, relevant to the problem of determining the ages of stars:

1. The higher the mass, the shorter the lifetime (calculated until the start of the final white dwarf stage or the supernova explosion) of a star, at fixed initial chemical composition. For example, a model of $1M_\odot$ with solar initial chemical composition will spend about 11 billion years in the MS phase, and will live for a total of about 12.7 billion years before becoming a white dwarf. The same numbers for a $3M_\odot$ model are 360 million years and 460 million

---

[12]Not only the main evolutionary properties just discussed but also other details of the evolution of the models depend on their initial mass. I still remember with pleasure the discussions I had with a very close friend, fellow student at university, bandmate, and then fellow astrophysicist, during several running sessions to prepare for road races. Before the stellar evolution exam, during our runs, we used to review various details of the physical and chemical evolution of stellar models and their dependence on mass. It was an excellent anaesthetic against training fatigue.

years, respectively. Stars that explode as core-collapse supernovae have total lifetimes typically on the order of million years.

2. The MS is the longest evolutionary phase. More advanced stages have increasingly shorter timescales. Also, the PMS phase is much shorter than the MS (it lasts roughly 1% of the MS lifetime).

3. The higher the initial metal mass fraction (keeping the helium mass fraction fixed) the lower the luminosity and T_{eff} of a star of a fixed initial mass during the major evolutionary phases, and the longer its lifetime.

4. The higher the initial helium mass fraction (keeping the metal content fixed) the higher the luminosity and T_{eff} of a star of a fixed mass during the major evolutionary phases, and the shorter its lifetime.

We should now pause to consider the issue of the nuclear reactions. Just to give a comparison in terms of human activities, the Sun releases in a second an amount of energy that corresponds to about 2 billion times the most powerful atomic bomb ever deployed on Earth, the so-called Tsar Bomba, detonated in the Soviet Union in 1961. The question that springs to mind is the following: How can the nuclear energy production in stars be so carefully tuned, to avoid an explosion or a gravitational collapse?

The answer is the strong dependence of the efficiency of nuclear reactions on temperature already mentioned in chapter 3, and the response of the gas pressure to changes of temperature (described by the equation of state of the gas, see the appendix). If for some reason the nuclear reaction rates raise above the level required to maintain hydrostatic equilibrium, the burning region heats up, the gas pressure rises, the gas locally expands against gravity, and this expansion, in turn, decreases the temperature of the burning core. This temperature decrease slows down 'immediately' the rate of nuclear reactions, because they strongly depend on the temperature, decreasing with decreasing T. If instead nuclear reaction rates become too low, the burning region cools, lowering the gas pressure. In this situation, gravity prevails and causes a contraction of the nuclear burning layers, leading to a local increase of temperature and a consequent 'immediate' increase of the reaction rates, to restore hydrostatic equilibrium.

It is also important to have an idea of how much mass is transformed into energy due to the nuclear reactions, taking the Sun as an example. According to the models, the Sun is in the MS phase, transforming hydrogen to helium within the inner 10% of its mass. Nuclear physics tells us that only 0.7% of the hydrogen mass is converted into energy by the fusion reactions, and at the end of the MS, when all hydrogen has been transformed to helium in the nuclear burning region, the Sun will have lost about just $0.0007 M_\odot$.

4.2.1 Binary Stars

Stellar evolution models show that the evolutionary pathway and lifespan of a star are set at birth by the value of its mass and, to a lesser extent, by its initial chemical composition. However, this is true only if stars evolve in isolation.

We know that a large fraction of stars in the disk of the Milky Way are in binary systems, meaning two stars orbiting around their centre of mass (see the discussion about eclipsing binary stars in the appendix)[13]. If one star is much more massive than the other one, the centre of mass may even be a point within the more massive component, like in case of the Solar system, where the centre of mass is inside the Sun, hence all planets orbit around it.

The evolution of stars in a binary system can be exactly like that of single stars, as long as they are distant enough. Their initial separation is crucial because the change of radius of the two components during their evolution (see next section) can lead to gas being stripped from one star and dumped onto the other one, or even lost into the surrounding space, depending on the initial distance of the two objects and their masses. When this happens, the binary system is an 'interacting binary' and the evolution of the two components can be radically different from the case of single stars. Interacting binaries, for example, explain the second broad class of supernovae we see in the universe, the so-called thermonuclear supernovae, as exploding white dwarfs[14]: The explosion destroys the white dwarf, and the ejected gas is made mainly

---

[13]There are also systems of more than two stars, although they are much rarer than binaries.

[14]These are the supernovae whose light curves – how their luminosity changes with time after the explosion – display a very regular behaviour, and have been used to determine distances to high-redshift galaxies. This has led to the discovery of the acceleration of the expansion of the universe, see chapters 1 and 2.

of iron. It is not yet well established what is the final configuration of the binary progenitor that produces this type of supernovae. It could be a system of two white dwarfs whose combined mass is larger than the Chandrasekhar mass, that merge and explode. Another possibility is a white dwarf that slowly accretes hydrogen from the surface of a companion, with the ignition of hydrogen-burning on the surface of the white dwarf that eventually triggers a thermonuclear explosion.

In this latter case, depending on how fast is the accretion, a fainter nova (hence the lack of the prefix 'super ') can be produced. Nuclear reactions ignite at the surface of the white dwarf, but only the accreted layers are ejected, leaving the white dwarf intact. This process of accretion and ejection can repeat in time, as long as the companion is dumping mass at the appropriate rate onto the white dwarf.

4.3 STELLAR EVOLUTION TRACKS AND ISOCHRONES

As discussed in the previous section, stellar evolution calculations predict that the internal chemical structure of the models changes with time. This is a response to the need to preserve hydrostatic equilibrium following the loss of energy from the surface. These changes have a direct effect on both the surface luminosity and the effective temperature of the models, two properties that can be readily compared with observations.

Figure 4.3 displays the time evolution of models for three different masses and the same initial chemical composition, in a modern version of the HRD, showing the surface luminosity on the vertical axis, and the corresponding effective temperature on the horizontal axis. I also show the lines of constant radius (in units of the solar radius $R_\odot$), taking advantage of the relationship between L, R, and T_{eff} discussed before. The evolution of stellar models in a HRD is called 'evolutionary track', or 'stellar evolution track', and these chosen masses are representative of the typical morphologies of the evolutionary tracks.

These three tracks display the evolution from birth (the moment a star reaches for the first time hydrostatic equilibrium) until just before the supernova explosion for the $20 M_\odot$ track, or the early stages of white dwarf evolution for the other two tracks. The $1 M_\odot$ track represents the evolution of the Sun, whose position in this diagram corresponds to the point that intersects the diagonal line marked with $1 R_\odot$. Typically, more massive models evolve at higher luminosities and higher T_{eff}, compared to less massive ones.

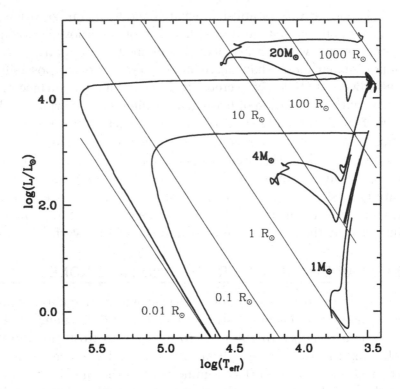

Figure 4.3 Evolutionary tracks of models with initial masses equal to 1, 4, and $20M_\odot$, with initial solar chemical composition [23, 78]. The diagonal lines of constant radius (in units of the solar radius $R_\odot$) are also shown. The vertical axis displays the logarithm of the ratio of the model surface luminosity over the luminosity of the Sun $L_\odot$ (a value equal to zero means that the model has the same luminosity of the Sun), while the horizontal axis displays the logarithm of the effective temperature T_{eff}, increasing towards the left of the diagram.

Moving along a track means to move in time, and it is striking to notice the large variations of T_{eff}, luminosity, and radius during the evolution of a model. The two horizontal sequences on the 1 and 4 $M_\odot$ tracks correspond to the transition from the AGB to the following white dwarf phase, with the large decrease in radius necessary to attain the small values (on the order of one hundredth of the solar radius, roughly equal to the Earth radius) typical of electron-degenerate objects.

The link between the position along these three evolutionary tracks and the internal structure of the models is shown in the Fig. 4.4. I will

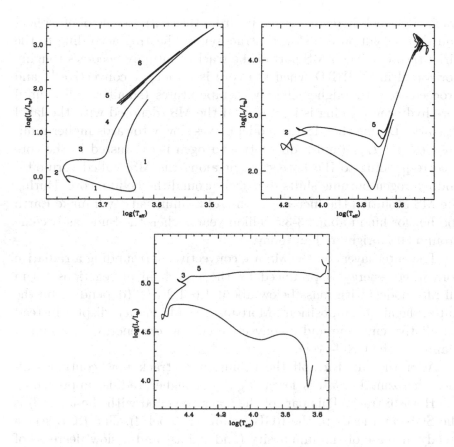

Figure 4.4 The tracks of Fig. 4.3 (corresponding to 1, 4, and 20 $M_\odot$ models, moving clockwise from the top-left panel) and the main evolutionary phases labelled in order of increasing age, as discussed in the text. The evolution of the 1 and 4 $M_\odot$ tracks is shown until the start of the transition to the white dwarf stage. In case of the 4 and $20 M_\odot$ tracks only the phases that display a morphology different from the 1 $M_\odot$ track are labelled. After the end of core He-burning the $20 M_\odot$ track basically terminates, because the following nuclear burnings in the core are very fast and the models hardly move in the HRD.

focus first on the 1 $M_\odot$ track in this figure. The evolution starts at the brightest point along the segment of the track labelled as *1*, the PMS, and the first, almost vertical part of this segment is called Hayashi track [15]. Models here are convective from the centre to the photosphere and their evolution is a slow contraction and heating according to the virial theorem. The PMS part of the track eventually becomes roughly horizontal in the HRD when the core is no longer convective[16], and proceeds towards higher effective temperatures (smaller radii), until core hydrogen-burning is ignited, and the MS (denoted with the label *2*) starts. During the MS the model moves slowly towards higher luminosities and T_{eff} (and radii) until hydrogen is exhausted in the core – corresponding to the hottest point along the MS, called turn-off – and hydrogen-burning shifts to a shell around the helium core. During the MS evolution the increase of the solar luminosity will make Earth too hot for life in about 3-3.5 billion years, when the Sun has become around 30% brighter than today.

The outer layers on the MS are convective, surrounding a radiative core where energy is produced by the p-p chain of reactions, as in all MS models with mass below about 1.0–1.2 $M_\odot$ (depending on the initial chemical composition). More massive MS models display instead a radiative envelope and a convective core where energy is produced mainly by the CNO-cycle.

After the MS turn-off the evolutionary track now continues almost horizontal towards lower T_{eff} and higher radius, approaching the Hayashi track. This part of the track (denoted with the label *3*) is the SGB, and leads to the RGB evolution (label *4*). The RGB sees a steady increase of the luminosity (and radius) and a slow decrease of T_{eff}, looking approximately vertical in the HRD[17]. The bottom edge of the convective region reaches increasingly deeper layers during the SGB and early RGB evolution, and throughout the whole RGB convection involves a large fraction of the mass, without reaching the inner hydrogen-burning shell.

In models with masses below about 2.0–2.3$M_\odot$ (but above 0.5$M_\odot$) the helium core becomes electron degenerate after the MS, but the temperatures to ignite helium-burning are nevertheless achieved at the brightest point on the RGB. The following core helium-burning evolution is labelled as *5*. The model luminosity drops fast after helium-burning ignition, then the track describes a sort of loop in the HRD

---

[15]Named after Chushiro Hayashi, who in 1961 predicted this vertical sequence of fully convective models in the HRD.

[16]This part of the PMS evolution is sometimes called Henyey track.

[17]During the RGB evolution the solar radius will reach approximately the size of the orbit of Earth.

and starts evolving again towards increasing L and gently decreasing T_{eff}. The exact value of the effective temperature of the model after helium ignition depends on how much mass has been lost from the surface during the RGB. Typical values are around 0.2 $M_\odot$ for models of about $1M_\odot$. The more mass is lost, the hotter the effective temperature after the ignition of helium-burning.

When both the luminosity starts to rise and T_{eff} to drop fast, helium in the core is exhausted, and the energy sources become helium and hydrogen shell burning, around an electron-degenerate core. The model enters the AGB phase (labelled as 6), with luminosity and radius steadily increasing until the whole envelope around the electron-degenerate C-O core is lost, due to very efficient mass-loss from the surface. This marks the end of the AGB phase, and the model continues its evolution as a white dwarf (more about white dwarfs in chapter 8).

The 4 $M_\odot$ track displays the same sequences discussed before, although the shape of the MS is quite different. For all masses with a convective core and a radiative envelope on the MS, the evolution is towards higher luminosities and lower effective temperatures, and the turn-off is marked by a sort of 'zig-zag' path, that leads to the SGB. The range of luminosities covered by the RGB evolution is narrower than for the 1 $M_\odot$ models (and the amount of mass lost in this phase is very small, because of the fast evolutionary speed along the RGB), a consequence of the fact that the helium core is not electron-degenerate. Also, the shape of the loop during the core helium-burning phase has a different morphology compared to the 1 $M_\odot$ track[18].

The 20$M_\odot$ track has a different morphology after the MS. The SGB phase is very short, and core helium burning starts when the model has still a high effective temperature. There is no RGB, nor AGB (no electron-degenerate core after the end of core helium-burning). The track essentially stops after the exhaustion of helium in the core, because all core burnings that follow are very fast (the total post core helium-burning lifetime is on the order of just 1000 years), and the models hardly move in the HRD.

---

[18]The 'wiggles' towards the tip of the AGB mark a very 'unsettled' phase towards the end of the AGB evolution, due to the so-called thermal pulses, whereby models experience fast variations of luminosity and effective temperature. This is common to all models that go through the AGB phase.

With stellar evolution tracks we can calculate the so-called isochrones, the main tool to determine the ages of stars. An isochrone[19] is the predicted HRD of a population of stars born at the same time with different masses and the same chemical composition, after an age t_i since formation. We will see in the next chapters how to use them.

The upper panel of Fig. 4.5 shows graphically the way to calculate an isochrone. The solid thick lines are three isochrones for ages equal to 100, 600 million years (million years will be denoted as Myr, that stands for Megayear), and 5 billion years (billion years will be denoted as Gyr, that stands for Gigayear). The dashed lines are evolutionary tracks of models with different initial masses, all with the same initial chemical composition.

An isochrone is simply the line that joins points belonging to the various tracks – one point along each track – where the age of the model is equal to the chosen t_i. It is a snapshot of the position in the HRD of coeval stars of different masses, at a given time after their birth. Different points along an isochrone correspond to the position in the HRD of stars with different initial masses at a fixed age, while points along an evolutionary track correspond to different ages for a fixed initial stellar mass.

Imagine attending the final of the 100-metre dash at the Olympics. There are eight participants, and eight cameras to film the race, each camera following just one participant. They are synchronized to begin filming when the race starts, and the recording stops when the runner crosses the finish line. As a spectator, I have a camera and take photographs (in burst mode, to have several snapshots per second) of the whole field, covering the track from the start to the finish line, at various instants during the race. All athletes will be in my shots, each of them at different spots along the track, because of their different speeds, their positions changing with time (from one shot to another) because they advance towards the finish line. Eventually, with time passing, the fastest athletes cross first the finish line and will be out of the field of view of my camera, one by one (I'm assuming here a perhaps unrealistic time resolution of my snapshots).

---

[19]This term comes from the Greek words 'chronos', that means 'time' and 'isos', or 'equal'.

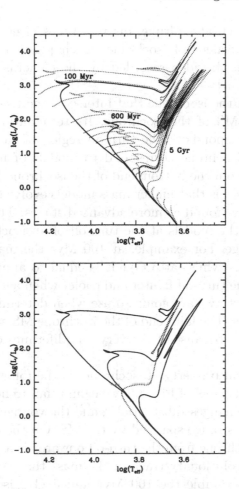

Figure 4.5 *Upper panel:* The dotted lines display evolutionary tracks of different masses [88], all with the same initial chemical composition (solar). Masses range from 0.5 (the faintest track) to 6 $M_\odot$. The tracks for the lowest four masses (0.5, 0.6, 0.7, and $0.8M_\odot$) have been calculated from the beginning of the MS only until the exhaustion of central hydrogen, because their MS lifetimes are much longer than the age of the universe. The evolution of the other tracks starts at the beginning of the MS and reach the end of the AGB phase. The thick solid lines are three isochrones (ages equal to 100 Myr, 600 Myr, and 5 Gyr) calculated from these tracks. Isochrones of increasing age display a shorter MS and a fainter turn-off. The PMS phase for all tracks is not shown because these masses have all reached the MS at the age of the isochrones displayed in this example. *Lower panel:* The same three isochrones of the top panel (solid lines) plus a 5 Gyr old isochrone calculated for a metallicity 20 times lower (dotted line).

My photos taken at multiple, discrete times after the start of the race, are the equivalent of isochrones, each photo being one single isochrone, while each video recorded by the cameras corresponds to an evolutionary track.

The section of the isochrone that intersects the tracks during their MS stage is the MS of the isochrone. Its termination is also called turn-off, and corresponds to the turn-off region of the parent tracks. Points along an isochrone correspond to models of increasing mass, when advancing from the bottom end of the isochrone MS towards the turn-off. The reason is that higher mass models evolve faster, hence for a fixed age they are found at more advanced stages. This also explains why the mass of the models at the turn-off of an isochrone decreases with increasing age. For example, at 100 Myr the mass at the turn-off is around $5M_\odot$, whilst at 5 Gyr it is equal to about $1.2M_\odot$. This decrease makes the turn-off fainter and cooler when age increases. This is a crucial property we are going to use when determining the age of stellar populations. The faint end of the isochrone MS seems essentially independent of age because of the very long lifetimes of the low mass models.

Another general property of isochrones is the following. If we look again at the upper panel of Fig. 4.5, we cannot fail to notice that along the MS the mass changes substantially, and the younger the isochrone, the larger the mass range spanned by the MS. On the other hand, the portion of the isochrone from the turn-off onwards is almost coincident with the single evolutionary track of the mass that is actually at the turn-off (see for example the 100 Myr and 5 Gyr isochrones in the figure). The reason is that the post-MS evolution of the models is much faster compared to the MS.

To explain this intuitively let's go back to the 100-metre dash race analogy. A long-lived evolutionary phase like the MS can be compared to a picture of a 90-metre stretch in the field of view. It takes the athletes about 9 seconds to cover the distance, hence any one of my shots (the isochrone) within 9 seconds from the start will display all athletes, albeit at different positions along the track. The post-MS evolutionary phases are roughly equivalent to a snapshot of a much shorter stretch of the track – let's say 2 metres or less – between the start and the finish line: Being very short, it is crossed almost instantaneously by the runners. At a given time, I will get in this shot only somebody whose speed is precisely the one that puts the athlete at that spot along the track at the time of my shot. Any other speed would place the runners out of the field of view of my camera.

Finally, the lower panel of Fig. 4.5 shows another important property that needs to be taken into account when determining the ages of stars. If we change the chemical composition of the models, the isochrones also change. This example shows that a lower initial metallicity shifts the isochrones to higher effective temperatures and overall higher luminosities, at a fixed age. Knowing the initial chemical composition of stars is therefore very important if we want to determine their ages with some accuracy.

In this respect, it is perhaps interesting to recall that the existence of stars with different initial chemical composition has been established only in the last 60–70 years. The famous spectroscopist Otto Struve concluded, in the summary of a 1950 International Astronomical Union symposium titled 'The abundances of the chemical elements in the universe', that the most striking result of the discussion was the remarkable degree of uniformity that has been observed in a large variety of astronomical sources. The symposium established a list of 'normal' abundances of the elements in the universe – the solar chemical abundance pattern – that seemed to constitute a universal law of nature.

The following year, Joseph Chamberlain and Lawrence Aller measured in a star belonging to the halo of the Milky Way an abundance of iron much smaller than solar. So entrenched was the idea of universal abundances that the authors themselves wrote in their paper how the observed Fe abundance appeared to be smaller than in the Sun, but their result had to be taken with caution [20]. Only in the 1960s the fact that the chemical composition of stars does not conform to a single, universal pattern, was finally widely accepted.

With the background and the tools at hand, we are now in the position to work out how to determine the ages of stars. But before doing so, we perhaps need a short digression, to convince ourselves that the theory of stellar evolution is indeed correct. At the end of the day, our results on stellar ages are based in a way or another on stellar evolution models and the 'cosmic clocks' they provide.

GLOSSARY

Isochrone: Hertzsprung-Russell diagram of stellar models with different masses, but the same initial chemical composition and age.

Stellar atmosphere: The low-density gas layers above the photosphere, where spectral lines form.

Stellar evolution track: Time evolution in the Hertzsprung-Russell diagram of models with a given initial mass and chemical composition.

Stellar model: A tabulation of physical quantities like temperature, pressure, density, but also abundances of chemical elements, or the rate of energy production, from the centre to the photosphere predicted for a star with a given mass and initial chemical composition.

FURTHER READING

J. Lequeux. *Birth, Evolution and Death of Stars*. World Scientific, 2013

R. Kippenhahn, A. Weigert and A. Weiss. *Stellar Structure and Evolution*. Springer, 2012

D. Prialnik. *An Introduction to the Theory of Stellar Evolution*. Cambridge University Press, 2009

S.G. Ryan and A.J. Norton. *Stellar Evolution and Nucleosynthesis*. Cambridge University Press, 2010

M. Salaris and S. Cassisi. *Evolution of Stars and Stellar Populations*. Wiley, 2005

CHAPTER 5

Great Hopes

As discussed in the previous chapter, calculations of stellar models have revealed how the structure (chemistry and physics) of stars is predicted to vary over time. For a fixed initial mass and chemical composition, models map the values of measurable quantities such as luminosity and effective temperature (or radius), onto their age. This gives us the opportunity to become forensic specialists, able to determine the age of stars from observations.

To this purpose, it is crucial to establish whether stellar models are reliable. We cannot see stars evolving across the HRD[1] and although we catch stars exploding, and observe stars which vary their luminosities more or less periodically, this is not a demonstration that stars evolve in time as predicted by our calculations. As a consequence, it is claimed by some outlets that the theory of stellar evolution is unproven and, depending on the personal opinion of the writer, it might even be totally wrong. If this is the case, the ages of stars derived with the methods presented in the next chapters may be meaningless.

I still recall vividly a meeting with a young Russian theoretical physicist about 30 years ago. We were having dinner at a nice restau-

---

[1]There is however the intriguing case of a star discovered in 1996 by the amateur astronomer Yukio Sakurai and named after him, located in the constellation of Sagittarius. Since its discovery, this star has first increased the luminosity sharply, the chemical composition of its surface has changed, then it has been enveloped by a thick cloud of dust and gas that blocks most of the emitted visible light. According to stellar evolution models, we are witnessing the real-time evolution of a newly formed white dwarf of about 0.6 $M_\odot$ through what is called a late thermal pulse, a violent ignition of helium-burning – a 'helium shell flash' – in the layers around the electron-degenerate C-O core, causing the ejection of some of the remaining envelope.

rant in a small town in central Italy, together with two Italian senior astrophysicists. While waiting for our food to be served, he began saying that he didn't believe stars were powered by nuclear reactions. He then went on describing his model of how stars work, but I have to admit I didn't understand much of what was being said, also because his theory was heavily based on magnetic fields, not one of my favourite subjects (meaning I wish I would know more about magnetic fields).

How do we prove our models are right? The theory of stellar evolution explains the observed positions of stars across the HRD in terms of evolutionary phases (stellar ages), masses, and initial chemical compositions, but this is not by itself a proof that the theory is correct. It is just the way these observations are interpreted in terms of our models.

One crucial feature of the theory of stellar evolution is that nuclear reactions are predicted to be efficient in the interiors of stars. And nuclear reactions, thanks to a large amount of energy per unit mass they can release[2], guarantee much longer lifetimes, compared to the case of just gravitational energy being available to power the stars. It is the interplay between gravitational contraction, efficiency of nuclear reactions, and onset of electron degeneracy in response to the loss of energy from the surface, that drives the changes in the internal structure, luminosity, and effective temperature of stars. How do we test that nuclear reactions are really at work in stars, and modify their chemical structure? Can we validate the internal physical and chemical stratification of our models?

5.1 RED GIANT BRANCH STARS

In 1967 Hans-Christoph Thomas, a then 31-year old astrophysicist working in Rudolph Kippenhahn group at the Max Planck Institute für Physik and Astrophysik in Munich (to be renamed Max Planck Institute for Astrophysics before the move to Garching), published a paper in the German-language journal 'Zeitschrift für Astrophysik' ('Astrophysics Magazine') [120]. He presented calculations of the evolution of models with 1.3 $M_\odot$, focusing mainly on the ignition of core helium-burning in the electron-degenerate core at the end of the RGB phase. Figures 1, 2, and 4 of the paper, related to the RGB evolution

---

[2]This is basically because nuclear reactions involve a change in an atom's nucleus, which is kept together by the strong interaction, the strongest of the fundamental forces of nature (see the appendix).

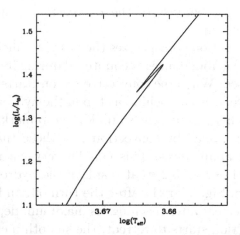

Figure 5.1 HRD of the RGB evolution of a $1M_\odot$ model with initial solar chemical composition.

before helium-burning ignition disclosed a new feature. The top panel of Fig. 5.1 displays this feature along the RGB evolutionary track of a slightly lower mass model $(1.0M_\odot)$.

As discussed briefly in the previous chapter, the evolution of the models along the RGB is usually towards increasing luminosities and gently decreasing T_{eff}. Thomas found that there is a point on the RGB (its location depends on the mass and initial chemical composition) at which this trend briefly reverses. We can identify this point at $\log(L/L_\odot)$ around 1.4 (luminosity about 25 times the solar luminosity) in Fig. 5.1. This behaviour is related to the evolution of the depth of the convective envelope above the helium core, and the nuclear reactions efficient during the previous MS. Along the SGB and during the early stages of RGB evolution, convection at the surface of the models reaches increasingly deeper layers, and when the lower boundary of the convective envelope gets too close to the hydrogen-burning shell, convection starts to slowly retreat toward the surface.

According to the calculations, convection at its maximum extension encroaches upon regions where the hydrogen abundance is slowly decreasing towards the centre, due to the nuclear reactions during the MS that have transformed some hydrogen into helium. Their efficiency decreases when the temperature T gets lower, and around the hydrogen-exhausted core, there are layers where the hydrogen abundance was only partially altered because of the decreasing temperatures when moving towards the surface.

Given that convection homogenizes the gas very efficiently, any variation of the chemical abundances encountered during its march towards inner layers is erased. When the convective region starts to retreat back towards the surface, it leaves behind a step in the hydrogen abundance, at the point of its maximum extension. This step marks the boundary between the layers mixed by convection and those that preserve the original hydrogen abundances. This can be very clearly appreciated in the upper panel of Fig. 5.2, that displays the hydrogen abundance profile of models in Fig. 5.1 right after the turn off, and when the convection starts to retreat after reaching its maximum depth. We can see that, when convection starts to retreat, the smooth increase of X (the hydrogen abundance) moving outwards from the H-exhausted core is broken by a step, followed by a constant value up to the surface. This constant value is slightly lower than the initial hydrogen abundance because layers with the pristine hydrogen have been mixed with layers with decreasing hydrogen content. We can also see in the figure that between the turn-off and the point of maximum depth of convection on the RGB, the hydrogen-exhausted region has increased its mass, because of the helium produced by the hydrogen-burning shell.

Let's now switch to the point of view of the hydrogen-burning shell, that is advancing towards the surface, adding fresh helium to the hydrogen-exhausted core. What is lying ahead of the shell is an amount of hydrogen smoothly increasing, and at this stage the surface luminosity of the models steadily increases. At some point, the shell encounters the sudden step-like jump in the hydrogen abundance, then followed by a constant abundance until the surface. When the shell crosses the abundance jump, the luminosity of the model decreases, to start increasing again once the shell has gone across the step.

This behaviour of the luminosity of RGB stars is a truly genuine prediction of the theory of stellar evolution. If confirmed, it would validate the evolution of the internal structure of the models from the MS to the RGB, and the presence of efficient nuclear burning, both in the core during the MS, and in a shell around the helium core during SGB and RGB. Thomas didn't discuss the observational implications of this result, this was done a year later by Iben [59], but an empirical confirmation had to wait about 20 years.

In the previous chapter, we have seen that the post-MS evolutionary phases of an isochrone overlap with the stellar evolution track of the

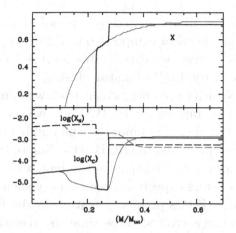

Figure 5.2 Internal chemical stratification of models in Fig. 5.1 right after the turn off (thin lines), and when the convection in the envelope starts to retreat after reaching its maximum depth (thick lines). The upper panel displays the hydrogen mass fraction X, the lower panel the mass fractions of C (X_C, solid lines), and N (X_N, dashed lines). The horizontal axis shows the distance from the centre parametrized in terms of the fraction of the total mass contained within a given point (M_{tot} is the total mass of the model): A value equal to zero corresponds to the centre, a value equal to unity denotes the photosphere.

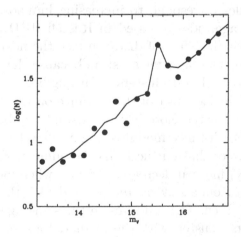

Figure 5.3 Luminosity function of RGB stars (filled circles) in the Milky way globular cluster M 3 [95] compared to a theoretical LF (solid line).

model at the isochrone turn-off. As we will see in the next chapter, a star cluster is very well approximated by a population of stars with the same age and initial chemical composition. Therefore, if the theory of stellar evolution is correct, we expect this feature to be present at a specific luminosity of the RGB of a star cluster.

But how to reveal its presence, given that we cannot observe the real-time evolution of stars on the RGB? Well, we can count the number of objects as a function of luminosity. The number of observed stars is expected to distribute across the RGB according to the speed of their evolution. The faster the evolution across a given luminosity range, the lower the number of stars expected, because they spend a shorter time at those luminosities. If this predicted feature in the HRD is real, the same luminosity range is crossed three times by the track, hence more stars are predicted to be there compared to the neighbouring luminosities. A plot of the number of stars counted in different luminosity intervals as a function of the mean luminosity of these intervals is called differential luminosity function, or just luminosity function (LF). Iben predicted the existence of a local enhancement of star counts in the LF of cluster RGBs at the luminosity of the feature first found by Thomas; in 1985 this prediction was finally confirmed by observations of the Milky Way globular cluster 47 Tuc [66].

Figure 5.3 displays the observed LF of RGB stars in M 3, another globular cluster of the Milky Way. The horizontal axis displays the apparent magnitude m_V of the stars, a proxy for their luminosity (decreasing magnitudes correspond to increasing luminosity), equivalent to the absolute magnitude M_V used in Russell HRD (see chapter 3) but uncorrected for the effect of distance (see the next chapter). The distance is virtually the same for all stars because the size of globular clusters (and almost all star clusters) is negligible compared to their distance from Earth; the effect of the distance on the brightness measured in magnitudes is therefore just a constant number, to be added to all the stars, the 'distance modulus' (again, see the next chapter).

We can clearly see that on the RGB the number of stars per magnitude (luminosity) interval decreases steadily when moving to higher luminosities, apart from a spike at m_V around 15.5, the local enhancement predicted by Iben. The solid line in the same figure is a theoretical LF for the same cluster, with the predicted magnitudes shifted to account for the cluster distance. The total number of observed RGB stars has been distributed across the RGB of an appropriate evolutionary track, according to the speed with which the model evolves. The mass of this evolutionary track follows from an estimate of the cluster age (see chapter 7), but the exact value is not critical.

The theoretical LF shows a striking agreement with observations, and displays a spike as observed, due to the evolutionary feature discovered by Thomas. Today, this spike in the LF is usually named RGB bump[3].

Another important result emerging from the comparison in Fig. 5.3 is the good match between the observed and predicted slope of the LF across the whole RGB. The slope is determined by the relative number of stars at different luminosities, that depends on how the speed of evolution changes across the RGB, as already mentioned. This agreement between theory and observations confirms the speed of RGB evolution predicted by the models.

Further empirical support for the internal structure of RGB models comes from measurements of chemical abundances. The upper panel of Fig. 5.2 shows that after convection has reached its maximum extension, the surface abundance of hydrogen has decreased compared to the initial value because layers with the pristine hydrogen have been mixed with inner layers with decreasing hydrogen content. The decrease of hydrogen is due to the production of helium at its expenses, therefore the surface helium abundance is predicted to have correspondingly increased. This phenomenon is called first dredge-up[4] and in principle could be tested by measuring the surface helium abundances along the SGB and RGB of star clusters. Unfortunately, no helium lines can be found in the spectra of RGB stars, because their photospheres are too cold. But in addition to He, also the surface abundances of carbon and nitrogen – that can be measured from spectra of RGB stars – are expected to change due to the first dredge-up. This can be seen in the lower panel of Fig. 5.2, that displays the corresponding C and N abundance profiles right after the turn-off, and after the first dredge-up.

In MS stars that burn hydrogen with the CNO-cycle, nuclear reactions that produce helium transform some C into N, and above a certain temperature also some O is converted into N (but the total

---

[3]This terminology makes me chuckle, because it reminds me of the clumsy Inspector Clouseau talking about 'A bump upon the head' of the maid Maria Gambrelli, in the classic Blake Edwards comedy 'A shot in the dark'. Despite this, I did a number of works together with my good friend and colleague Santi Cassisi and other collaborators, on the RGB bump as a test of stellar evolution models.

[4]Qualitatively similar convective mixing phenomena also happen during the AGB evolution and are called, not surprisingly, second- and third dredge-up.

number of C+N+O stays constant). Even in models where energy is produced mainly by the p-p chain, the CNO-cycle is still able to modify the abundances of C and N in the hydrogen-burning region.

Convection at its maximum extension engulfs layers where C has been depleted and N enhanced, and this causes a decrease of the surface abundance of C and an increase of N after convection starts to retreat, as shown in Fig. 5.2. Iben described this phenomenon in 1967 [58], and the first observational confirmation of this prediction came two years later [43]. Measurements of C and N abundances along the SGB and RGB of star clusters show very clearly the effect of the first dredge-up on the surface abundances of C and N, another crucial test of stellar evolution models.

5.2 NEUTRINOS FROM THE SUN

According to the theory of stellar evolution, nuclear reactions in the Sun are transforming four hydrogen nuclei (four protons) into a helium nucleus (made of 2 protons and 2 neutrons) two positrons (positively charged electrons, see the appendix) and two neutrinos. There are three types of neutrinos in nature, the electron neutrino, the tau neutrino, and the muon neutrino (see the appendix): Solar neutrinos are predicted to be electron neutrinos. They easily escape from the Sun, and their detection here on Earth would be yet another decisive and direct proof that nuclear reactions are at work in stars.

In 1964, Raymond Davis and John Bahcall proposed an experiment to detect the neutrinos produced by nuclear reactions in the Sun, by measuring the number of radioactive argon atoms that neutrinos would produce when crossing a tank of cleaning fluid (mainly chlorine), approximately the size of a swimming pool[5]. The experiment was built in the Homestake Gold Mine in Lead, South Dakota, in the United States[6]. Theoretical models of the Sun[7] predict almost a hundred

---

[5] Bruno Pontecorvo proposed the use of this reaction to detect neutrinos in a 1946 paper.

[6]The underground location minimizes the 'noise' from cosmic rays. Cosmic rays are high-energy electrically charged particles, mostly protons and helium nuclei, coming from the Sun, novae, supernovae: When they hit the nuclei of atoms in the upper Earth atmosphere, more particles are created, and these can produce unwanted noise in the neutrino detector. Putting the experiment in a mine minimizes the probability that these unwanted particles reach the detector, because they cannot penetrate the ground as easily as neutrinos do.

[7]More on the calculation of solar models in chapter 7.

billion neutrinos crossing an area of one centimetre square of our skin every second, but only a few argon atoms would be produced per week because neutrinos hardly interact with other particles. That's also why we don't feel the effect of the large number of neutrinos crossing our bodies.

The experiment was performed by Davis, and the results announced in 1968: Neutrinos were indeed detected, thus confirming the hypothesis of nuclear energy generation in stars. However, their number was only one-third of what predicted by models for the Sun. Over the next thirty years, four more experiments (in Japan, Russia, and Italy) confirmed the detection of solar neutrinos, but again their number was a factor 2–3 lower than predicted by models. Three main possibilities to solve this problem were put forward.

The first obvious one was some error in the results of the experiments. The second possibility was that solar models were wrong, with pressure and temperature in the core of the Sun being different from the model predictions. Perhaps one of the most unusual solutions was proposed by Stephen Hawking, who imagined a low-mass black hole being nested inside the core of the Sun [47]. He wrote that a black hole with a mass of 10^{14} Kg, much less than the mass of the Sun (about 10^{30} Kg), might have an effect on the small central region where energy is generated and could be the reason why the observed flux of neutrinos is not as predicted.

I remember very clearly this 'solar neutrino problem' when studying stellar evolution during my undergraduate studies at 'La Sapienza' University in Rome, in the late 1980s. Vittorio Castellani, who was delivering the course on stellar astrophysics, was working on this problem and discussed it extensively during his lectures and in his textbook. Somehow I was not too bothered, like I guess it was the case of most of the stellar astrophysics community. It seemed inconceivable to me that models of the Sun could be massively wrong.

Another possibility to explain the solar neutrino problem was related to the physics of neutrinos, and not to stellar astrophysics. As early as 1969 Vladimir Gribov and Bruno Pontecorvo proposed a solution of the discrepancy found by Davis' experiment, based on the concept that neutrinos oscillate between different types. This was an idea being developed by Pontecorvo[8] independently of the solar neutrino

---

[8]Pontecorvo had an 'eventful' life. He studied physics at my same university 'La Sapienza' in Rome under Enrico Fermi, and worked in his group. In 1934 he moved

problem. Solar models predict the production of just electron neutrinos, and the experiment was based on the detection of only this type of neutrinos; but if they oscillate between different types during their travel from the solar core to Earth, our measurements will obviously provide fewer neutrinos than expected. The oscillation of neutrinos also requires that they have a mass different from zero, contrary to the Standard Model of particle physics, which prescribes that neutrinos are massless.

Indeed neutrino oscillations are the culprit, as demonstrated by the Sudbury Neutrino Observatory (SNO), located in a mine in Sudbury, Canada. The measurements at SNO were sensitive to the electron neutrinos, and to the total number of neutrinos of all types. The result was clear: The total number of solar neutrinos is as predicted, but the number of electron neutrinos is too low, because electron neutrinos transform into tau and muon neutrinos during their travel from the Sun to Earth. The lack of sensitivity to the muon and tau neutrinos is the reason why earlier experiments measured such a deficit of neutrinos. Neutrino oscillations have also been detected from experiments with particle accelerators, and with nuclear power plants.

Nuclear reactions are happening in the Sun precisely as predicted by stellar models.

5.3 NEUTRINOS FROM SUPERNOVAE

On 22 February 1987 the star Sanduleak −69o202 looked like a normal massive star, evolving slowly but surely towards its unavoidable final

---

to France, where he joined the French Communist Party, and during the Second World War moved to the United States. In 1949 Pontecorvo was working in Britain for the Atomic Energy Research Establishment, near Harwell, but after his German colleague Klaus Fuchs was arrested for espionage in February 1950, he was considered a security risk (although there was no proof he was a Soviet spy) and was offered another job at the University of Liverpool (he wasn't impressed when he visited, and neither was the vice-chancellor of the university, because of Pontecorvo poor English) where he could not have access to top-secret material. He eventually accepted the offer in July, saying that he would start at first part-time in October, but while on holiday in Italy in summer 1950, at the age of 37, he defected to the Soviet Union. Nobody heard of him again in the West until he gave a press conference in Moscow in 1955, to explain that he had defected to the Soviet Union because in his view the West was intent on an atomic war to achieve world domination. While working in the Soviet Union Pontecorvo was allowed his first trip abroad only in 1978, to Italy. I met him twice when I was studying at 'La Sapienza': He was friendly and charming, but also looked very frail. He died only a few years later.

destiny as core-collapse supernova. It was visible in the LMC, at a distance of about 160,000 light-years.

The day after, images of the galaxy did not show anything unusual, but on February 24 Ian Shelton[9] and independently Oscar Duhalde[10] working at different telescopes at the Las Campanas Observatory in the Southern Atacama desert in Chile – and later in the day Albert Jones, an amateur astronomer in New Zealand, the most prolific variable star observer of all time – discovered a new bright object in the LMC.

A few days later it was identified as the star Sanduleak −69o202, which had become the brightest and only naked-eye supernova (named Supernova 87A) since 1604 Kepler supernova in the Milky Way[11].

According to the models, in the last frantic instants before the explosion, a neutron star forms in the iron core, a process that releases a large flux of neutrinos: Protons in the core 'absorb' free electrons and transform into neutrons releasing neutrinos. These neutrinos reached Earth hours before any visible sign of the explosion, because they were released promptly at the formation of the neutron star, while the ejection of the mantle around the iron core takes several hours longer. Over 13 seconds, starting at 07:35 and 35 seconds Greenwich Mean Time on 23 February, 11 of these neutrinos were detected by the 'Kamiokande' detector in Japan, while 8 neutrinos were recorded by a detector in Brookhaven in the US, the start of the events in the two detectors being essentially simultaneous[12]. These detections were discovered days after the announcement of the new supernova in the LMC, by analysing the data of the previous days.

The duration of the event and the energy of the neutrinos were in agreement with the predictions of the formation of the neutron star right before the supernova explosion, another confirmation of the stellar models, even in such advanced and fast-evolving stages.

---

[9]Because of technical problems, he stopped his work for the night, and processed some test images of the LMC, noticing a new bright star that wasn't there before.

[10]While taking a break from his work he looked outside and saw a new object in the LMC, visible to the naked eye. He had to go back to continue work and temporarily forgot about the new object.

[11]This was very likely a thermonuclear supernova, named after Johannes Kepler, who described it in his book *De Stella Nova* (On the new star).

[12]If the neutron star had formed only a few minutes later, the neutrino signal would have been lost by Kamiokande, because of a two minutes blank period due to scheduled system maintenance. Masatoshi Koshiba, Professor Emeritus at the University of Tokyo, would have missed his 2002 Nobel Prize awarded for this discovery.

5.4 HELIOSEISMOLOGY: THE SUN IS IN TUNE

As we have seen, a star consists of gas held together by gravity in hydrostatic equilibrium. However, a spherical shell at a distance r from the centre of a star, is not static like in case of a solid sphere, rather it is subject to localized random motions (like for any gas) across its volume, both outwards and inwards, that alter its shape on very short timescales. The displacements are small, and their random nature guarantees that on average the shell stays spherical, at the distance prescribed by the condition of hydrostatic equilibrium.

The situation is different when there is a mechanism in place to sustain a more regular pattern of oscillations. Let's consider for example the case of hitting the skin of a drumhead with a drumstick[13]. Depending on where exactly I hit the drumhead and the type of drumstick (a standard drumstick with a small head, or a mallet with a larger end, which I use to play some of the psychedelic Pink Floyd songs) different modes of vibrations are excited. For example, in the so-called fundamental mode the whole skin is displaced downwards, and then upwards during the oscillation, while in other more complex modes different sections of the skin are displaced in different directions at the same time, either downwards or upwards. If I hit the drumhead only once, the oscillations die out fast, but if I keep hitting the same place on the drumhead with the same strength, oscillations will be maintained.

Something similar happens in stars. A gas sphere will oscillate (pulsate) in many different modes when suitably excited, and the frequencies of these oscillations depend on the sound speed across the star, which in turn depends on pressure, temperature, and properties related to whether the stellar matter is a perfect gas or deviates from this idealized state (which are in turn affected by the gas chemical composition).

The best-known examples of pulsating stars are the so-called RR Lyrae (firstly discovered at the end of the nineteenth century) and Cepheid (discovered at the end of the eigtheenth century) variables. According to stellar evolution models these stars are currently evolving in the core helium-burning phase: They periodically increase and decrease radius (and luminosity), maintaining a spherical shape throughout the pulsation cycle, hence they are classified as 'radial pulsators'. The

---

[13]I can guarantee that when I play my drums I don't really think of the oscillation modes of the skin on my toms or the snare drum. I might be more concerned about coordinating my right leg with the right hand when playing the Purdie shuffle.

pulsation affects only the external layers and does not reach the inner energy generation region. RR Lyrae stars are found in old populations, have masses typically equal to 0.6–$0.7 M_\odot$ and periods of a fraction of a day. Cepheids have periods ranging from a few days up to roughly 100 days, and masses from a few to a few tens of solar masses, hence they are brighter than RR Lyrae stars. They also follow a well-defined relationship between the period of pulsation and average luminosity, as discovered by Henrietta Leavitt in 1907, when studying Cepheids in the nearby Small Magellanic Cloud. This period-luminosity relationship of Cepheids was employed by Hubble to determine the distances[14] that led to his announcement of the distance-redshift relationship for galaxies (see chapter 2).

What keeps the pulsation going (in the same way as I continue hitting my snare drum in the drumhead example) is the ability of the outer layers of these stars – due to the properties of the opacity of the gas – to absorb, accumulate, and then release the energy gained during the contraction phase of the pulsation. This mechanism to sustain the pulsation is efficient only in stars within a well-defined region of the HRD, the 'instability strip'[15].

There are however other possible pulsation modes in stars, like when different sections of the drumhead move in different directions during the oscillations. These more complex modes cause some regions across the surface to expand while others contract at any given time during the pulsation cycle, and are called non-radial pulsations[16], because technically the spherical shape of the star is not preserved, even though the amplitude of these pulsations is totally negligible compared to the size of the star. While in the case of radial pulsations the variation of the optical brightness during a pulsation cycle is on the order of tenths of a magnitude up to a magnitude, non-radial pulsations cause variations on the order of micro-magnitudes (10^{-6} magnitudes).

In the Sun up to one million of these non-radial modes are excited by the raising and falling gas bubbles within its convective envelope,

---

[14]Measurements of Cepheids periods in a galaxy would provide the intrinsic average luminosity of these stars, hence the distance when compared to their observed average brightness.

[15]I was invited a few years ago to a very interesting conference in Poland, about RR Lyrae and Cepheid stars. I'll never forget the venue of the conference dinner: The Jan Haluszka Chamber, carved in green rock salt 135 metres underground, in the Wieliczka Salt Mine.

[16] Chaim Pekeris and Thomas Cowling published major theoretical works on non-radial pulsations of stars in the late 1930s and early 1940s.

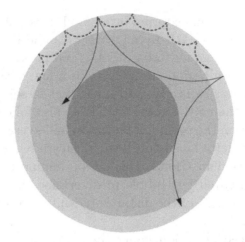

Figure 5.4 Diagram showing the propagation of sound waves (represented by the solid and dashed lines) in a cross section of the interior of the Sun (courtesy of A. Miglio).

that travel at speeds close to the local speed of sound. Their movement creates a spectrum of sound waves – regions of compression and rarefactions of the gas which travel across the star – like the sounds we hear when water is boiling in our pot on the stove. They propagate between the surface (where they cause the observed non-radial pulsations) and an inner boundary, as shown qualitatively in Fig. 5.4, and their speed depends on how often the gas particles collide. More collisions mean that the sound wave propagates faster, hence the speed of sound increases when the density is higher at fixed T, or the temperature is higher (faster particles) at fixed density. Also, lighter elements propagate the sound faster at a given temperature, because the particles move faster and have more collisions.

The lines in Fig. 5.4 trace the path of sound waves of different frequencies, bounded at the top by the steep decrease of the density of the gas near the photosphere, and at the bottom by the increasing sound speed that bends the path of the waves travelling inwards, until they are eventually refracted back towards the surface. Different sound waves thus probe different depths inside the Sun, and the distribution of the periods of the non-radial oscillations they cause – that depend on the depth reached by the sound waves, and the sound speed of the gas layers they cross – provide a powerful diagnostic of the internal structure of our star.

This is the way geophysicists study the interior of our planet. Seismic waves generated by earthquakes, volcanic eruptions, landslides, or man-made explosions travel through the Earth's layers, propagating with a path and a velocity that depend on the internal structure of the planet. They are recorded by networks of seismographs distributed around the world, and the study of their propagation discloses details of the Earth interior. This is how we have discovered that the outer part of the Earth's core is liquid.

In 1962 Robert Leighton, Robert Noyes, and George Simon observed for the first time oscillations of the surface of the Sun, with a period around 5 minutes [70]. They measured tiny oscillations of different points across the solar disk by taking advantage of the Doppler effect to detect motions of the surface. Following further observations, by the mid-1970s it was clear that there were multiple modes of oscillation with a range of periods between 3 and 15 minutes, and the largest amplitudes around 5 minutes. Jørgen Christensen-Dalsgaard and Douglas Gough[17] suggested the use of the frequencies of these oscillations to determine the internal structure of the Sun [24], marking the beginning of the field of 'helioseismology' (a term coined by Gough), the study of the structure of the Sun through its oscillations. 'Asteroseismology' denotes the same method of investigation when applied to stars other than the Sun..

The top panel of Fig 5.5 displays the internal structure of a theoretical model of the Sun (a solar model), showing the logarithm of pressure (in Pascal)[18], temperature (in Kelvin), and density (in kilograms per cubic metre), as a function of the distance from the centre. The same panel shows also the values of the hydrogen and helium mass fraction across the model: The hydrogen abundance starts to decrease due to

---

[17]I had the pleasure to meet both of them several times. In 2007 I was invited to a conference in honour of Douglas Gough in Cambridge, where I learned the meaning of the word 'high-table', during the conference dinner (I found myself sitting at what I was told was the high-table). During my very pleasant visit at Jørgen Christensen-Dalsgaard institute in Aarhus (Denmark), I discovered the most amazingly space-efficient hotel room I have ever been in. A miracle of compactification that would make a string theorist proud.

[18]A more familiar unit of measure of pressure used in everyday life is the 'bar'. As an example, typical values for a car tyre's pressure are slightly more than 2 bar. Given that 1 bar is equivalent to 100,000 Pascal, the pressure at the centre of the Sun predicted by the solar model is about 100 billion times the pressure of a car tyre.

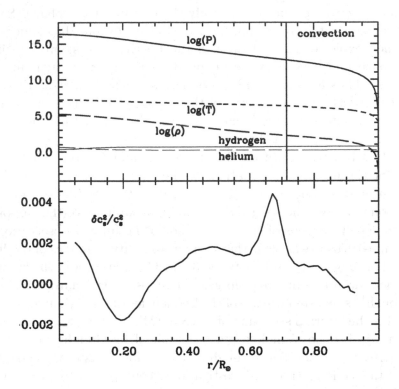

Figure 5.5 *Upper panel:* Values of hydrogen and helium mass fractions, and the logarithm of pressure (in Pascal), temperature (in Kelvin), and density (in kilograms per cubic metre), as a function of the distance r from the centre of a solar model normalized to the total radius. The vertical solid line marks the position of the lower boundary of the solar convective envelope. *Lower panel:* Difference between the observed and model values of the square of the sound speed (c_s^2) divided by the model value, across the interior of the Sun as determined from helioseismology [38].

the nuclear reactions (and the helium abundance increases) when the distance from the centre decreases below about 30% of the solar radius.

The bottom panel of the same figure displays the ratio of the difference between the observed and predicted values of the square of the sound speed (c_s^2) to the model values (that depend on pressure, density, temperature and chemical composition), as a function of the distance from the centre of the Sun[19]. Observations do not probe the very centre of the Sun, but they reach well into layers where nuclear reactions are efficient to convert hydrogen to helium. The agreement between theory and observations is impressive. Differences between observed and measured c_s^2 are within less than 0.5%. The energy generation region sampled by these observations is also in good agreement with the model, once again confirming the efficiency of nuclear reactions in stars[20].

5.4.1 Asteroseismology

Stellar evolution models predict that stars with an initial mass between 0.5 and about $2.3M_\odot$ ignite helium-burning in an electron-degenerate helium core: After ignition, the matter in the core eventually starts to behave again approximately like a perfect gas. As a consequence of these different physical conditions in the core, RGB models in this mass range have higher central densities compared to their counterparts with roughly the same mass, but in the core helium-burning phase. This prediction can now be tested with asteroseismology.

The mechanism that causes the non-radial oscillations of the Sun is expected to be at work also in RGB stars and their core helium-burning progeny with mass above 0.7–$0.8M_\odot$, that are all predicted to have extended convective envelopes. These solar-like oscillations were first observed in stars other than the Sun with the space-based Microvariability and Oscillations of STars telescope (MOST – active in the years 2003–2014). An increasingly larger number of detections has followed, thanks to the satellites Convection, Rotation and planetary Transits (CoRoT – active in the years 2006–2012), *Kepler* (named after Johannes Kepler– active in the years 2009–2018), and the ongoing Transiting Exoplanet Survey Satellite (TESS – 2018-ongoing) mission.

---

[19]Models of the Sun, as well as stellar models calculated to study populations older than around 1 billion years (isochrones older than this age are populated by masses lower than about $2M_\odot$) do not usually include the effect of rotation. The typically 'slow' rotation measured in samples of stars in this age and mass ranges is not predicted to have a major effect on their structure and lifetime. The inclusion of rotation is much more important in models for younger stellar populations.

[20]There is an ongoing debate regarding the metallicity of the Sun. Models calculated with recent determinations of the solar metallicity show a worse agreement between observed and predicted values of c_s^2 compared to Fig. 5.5. This is especially true in the regions right below the boundary of the convective envelope, however the largest differences are still at most on the order of just 1%.

These space-based telescopes (typically geared toward the discovery of planets that transit in front of the target star) are able to detect patterns of minute variations of stellar brightness[21], whose periods and amplitudes can disclose the presence of non-radial pulsations, even though the actual stellar disks are not resolved (the target stars are still point-like objects).

The predicted change of the internal density profile (how the density changes across the stellar model) between RGB and core helium-burning models leaves a signature in the periods of the non-radial oscillations of these stars, that has been recently detected thanks to observations with the *Kepler* satellite [8]. This provides yet another successful confirmation of the theory of stellar evolution.

GLOSSARY

Asteroseismology: Study of the structure of stars other than the Sun, through their oscillations.

Dredge-up: Mixing of material that has undergone nuclear fusion into the outer layers of stars with deep convective envelopes.

Helioseismology: Study of the structure of the Sun through its oscillations.

Luminosity function (differential): Trend with luminosity of the number of stars observed per luminosity interval.

Neutrino oscillations: Quantum mechanical phenomenon whereby a neutrino created with a specific 'flavour' (electron, muon, or tau) is later measured to have a different flavour.

Non-radial pulsations: Oscillations of a star surface around the hydrostatic equilibrium radius, with some parts of the surface moving inwards, while others move outwards at any given time.

Radial pulsations: Oscillations of a star surface that preserve its spherical shape (the whole surface expands or contracts uniformly at any time).

---

[21]For example, CoRoT could detect brightness variations with an accuracy of about 20 parts per million.

FURTHER READING

C. Aerts, J., Christensen-Dalsgaard and D.W. Kurtz. *Asteroseismology*. Springer, 2010

A K. Mann. *Shadow of a Star: The Neutrino Story of Supernova 1987A*. W H Freeman & Co, 1997

M. Stix. *The Sun. An Introduction*. Springer, 1989

CHAPTER 6

Hipparcos' Legacy

We have seen in chapter 4 how stellar evolution models predict that stars of all initial masses and chemical compositions experience large variations in effective temperature, luminosity, and radius during their evolution. The same models link the actual values of these quantities to the time elapsed since the formation of the stars, paving the way for the determination of stellar ages through measurements of their observable properties.

In this chapter, I discuss the connection between these age-sensitive stellar properties and observations. In other words, I will present the translation from the language of theoretical models to the language of observations.

6.1   MAGNITUDES AND COLOURS

For the large majority of stars we can only measure proxies for L and T_{eff}; they are the 'magnitude' (in a given wavelength range) and 'colour' (or 'colour index' the difference of the magnitudes in two different wavelength ranges), respectively. The definition of magnitude will probably look a bit convoluted, a reflection of the historical development of astronomy, harking back to a time when stars were unknown (and considered to be unknowable) objects, as I am going to show you now.

The first quantitative measures of the brightness of stars can be traced back to the Greek astronomer Hipparcos who, around 150 BC, produced a catalogue of 850 stars with positions in the sky and a ranking according to their apparent brightness. In his system, the brightest stars were assigned a 'magnitude' equal to 1, the next brightest stars magnitude 2 and so on to the faintest stars, just visible to the naked eye (he did not have telescopes at the time) which were assigned magnitude 6. These numbers had obviously nothing to do with the intrinsic luminosity of the stars but ordered them according to their apparent brightness.

The discovery of fainter stars after the invention of telescopes in the early seventeenth century, followed by the development of instruments to measure the number of photons received from the target stars, pushed the astronomers to quantify, standardize, and extend Hipparcos magnitude system, rather than to replace it. In 1856 Norman Pogson proposed that a star of magnitude 1 is 100 times brighter than a star of magnitude 6. This brightness difference, coupled to the general belief that human eyes detect differences in light intensity logarithmically, led to the following relation between the observed magnitudes (m) of two stars (1 and 2) and their apparent brightness (f – energy per second and per square metre registered with our detector): $m_1 - m_2 = -2.5 \log(f_1/f_2)$. The reference comparison star for this relative brightness scale is usually assumed to be Vega (it would be star 2 in this example) the brightest star of the Lyra constellation, in the Northern hemisphere; Vega is typically assigned an observed magnitude (named apparent magnitude by the astronomers) equal to zero, and magnitudes of stars decrease with increasing measured brightness, becoming negative for objects that look brighter than Vega[1].

With our present technology, it is not possible to build one instrument able to efficiently survey the entire electromagnetic spectrum. Moreover, the Earth atmosphere reduces the brightness of astronomical objects due to scattering and absorption of molecules, particularly at wavelengths above 1000 nm (the infrared range, although there are wavelength 'windows' in the infrared that are transparent to photons) and in the ultraviolet below about 300 nm. Space telescopes in orbit above the atmosphere are necessary to detect ultraviolet and X-ray (wavelengths even shorter than the ultraviolet part of the spectrum) emission from stars and celestial bodies in general, and also photons in several infrared wavelength ranges.

Magnitudes of stars observed with a given instrument are therefore not the so-called bolometric magnitudes, obtained measuring the stellar brightness over the entire wavelength range covered by its blackbody emission. Instead, observations are standardized by using sets of filters

---

[1]For example, a star half as bright as Vega has $m_1 = -2.5\log(0.5)=0.75$, while a star twice as bright as Vega has $m_1 = -2.5\log(2.0)=-0.75$.

('photometric filters') that let through to the detector only photons in well-defined wavelength 'windows'. They are the equivalent of the common photographic filters, pieces of glass or resin usually placed in front of a camera's lens, which are used for example to allow only certain colours coming into the camera.

A classic and much employed photometric system is the Johnson-Cousins system, comprising the filters U, B, V, R, I, each one covering a range of about 100–200 nm. The U filter is centred at wavelengths around 350 nm, the B filter around 450 nm, the V filter around 550 nm and corresponds roughly to the optical wavelength range (the wavelength range of light visible to the human eye), the R filter around 650 nm, and the I filter around 800 nm, in the near-infrared. Some of the most used filters available on board the Hubble Space Telescope are also very close to the Johnson-Cousins ones.

The reference point of the apparent magnitudes in most photometric filters is still Vega, whose apparent magnitude is usually set to $m=0$ (or close to zero) in every filter[2]. This means that, by definition, Vega has a magnitude equal to zero at all wavelengths, hence a generic star will have different magnitudes in different wavelength ranges unless its spectrum is identical to that of Vega. With this choice of the zero point, in the V filter the Sun has an apparent magnitude $m_V = -26.7$, Sirius has $m_V = -1.5$ and Proxima Centauri, the closest star to the Sun, has $m_V = 11.1$.

Obviously, differences of measured apparent magnitudes do not reflect differences of the intrinsic stellar brightness, for the measured brightness of an object scales as the inverse of the square of its distance. Two stars with the same intrinsic luminosity but at different distances will have different apparent magnitudes.

In addition to the effect of distance, the apparent magnitude is also affected by the so-called interstellar extinction[3]. This is the dimming of distant objects due to the dust present in the interstellar medium along the line of sight. The Sun is located in the disk of the Milky Way, which contains copious amounts of interstellar medium out of which new stars are born. This is made of about 99% gas (mainly hydrogen) and 1% dust, in the form of small grains of silicates, iron, carbon,

---

[2]Recently astronomers have also started using the so-called AB-magnitudes. In this system, the reference brightness is that of a fictitious object having at all wavelengths the same brightness that Vega has at a wavelength of 550 nm.

[3]The existence of interstellar extinction was first established by Robert Trumpler in the 1930s.

frozen water, and ammonia ice. Their typical size is comparable to the wavelength of light in the blue and ultraviolet part of the spectrum. Photons travelling in our direction with especially these wavelengths are very efficiently scattered or absorbed and reemitted isotropically by the dust. As a result, they tend to be removed from the beam of light that reaches our telescopes, and this makes the star fainter, especially at blue and shorter wavelengths. The effect of this so-called extinction (denoted with A) is quantified in terms of magnitudes, as -2.5 times the logarithm of the ratio of the measured brightness to the brightness that would be recorded in case of no extinction. With this definition, A is always a positive number[4]. Empirical relationships among the extinction A in different photometric filters do exist: For a fixed value of the extinction in V (A_V), the corresponding extinction in B (A_B) is about 1.3 times larger, whilst the extinction in I (A_I) is about 60% of A_V, for example.

If the extinction along the line of sight is different from zero, the apparent magnitude m_V of a star (using the V filter as an example) can be written as $m_V = m_{V,0} + A_V$, where $m_{V,0}$ is the extinction-free value (the suffix 0 usually denotes extinction-free quantities). If A_V is known, we can obtain $m_{V,0}$, and if the distance is known we can calculate the so-called absolute magnitude (M_V), that is the extinction-free magnitude the star would have if it were at a distance of 10 pc from us. The difference between apparent (corrected for extinction) and absolute magnitude is called 'distance modulus' and is equal to $(m - M)_{V,0} = 5 \log(D/10)$, where D is the distance in parsec[5]. The absolute magnitude (in V in this example) of a star is therefore given by $M_V = m_V - A_V - (m - M)_{V,0}$. We often also use the quantity $(m - M)_V = (m - M)_{V,0} + A_V$ called 'apparent distance modulus', that is the sum of the effect of both distance and extinction on the measured stellar magnitudes; we can then write M_V also as $M_V = m_V - (m - M)_V$.

It is the absolute magnitude that is related to the intrinsic brightness of a star, because it is calculated placing the target at a fixed

---

[4]The measured brightness is always lower or at best the same as the case of no extinction. The logarithm of a number less than unity is always negative, and the minus sign in front of the logarithm makes A positive, or equal to zero if the brightness is unaffected by extinction.

[5]Distance moduli can be both positive (distances larger than 10 pc) and negative (distances shorter than 10 pc). A distance modulus equal to zero corresponds to 10 pc, $(m - M)_{V,0} = 1$ is equal to D=15.8 pc, while $(m - M)_{V,0} = -1$ means D=6.3 pc.

distance (and the effect of extinction has been corrected for), the same for all stars. The Sun has an absolute magnitude in the V filter equal to M_V=4.8, Sirius has M_V=1.4, and Proxima Centauri M_V=15.6, meaning that Sirius is about 23 times brighter than the Sun in this filter and the Sun is in turn about 21,000 times brighter than Proxima Centauri. In a shorter wavelength filter like U, Sirius is instead about 125 times brighter than the Sun.

To understand better why the absolute magnitudes of stars depend on the photometric filter, let's recall that measurements of how their brightness vary when we observe photons of different wavelengths – after correcting for the effect of interstellar extinction – produce stellar spectra, characteristics sets of curves in brightness versus wavelength diagrams. Crucially, these curves have almost precisely (apart from the presence of spectral lines that are absent in blackbody spectra) the shape of what we would measure when observing the radiation inside ovens with a tiny hole in their side, and varying temperatures.

This radiation is the empirical counterpart of a 'blackbody', an idealized cavity in which photons are in thermal equilibrium with the walls, meaning that, on average, the walls emit as many photons as they absorb. Theoretical physics tells us that a blackbody spectrum depends only on temperature, the hotter the blackbody, the more light it gives off at all wavelengths (the Stefan-Boltzmann law), and the shorter the wavelength of its peak brightness (the Wien law), as shown in Fig. 6.1.

For example, the solar spectrum peaks at about 500 nm – within the wavelength range of the colour yellow – corresponding to a temperature of about 5800 K, according to the blackbody laws[6]. A blackbody twice as hot as the Sun has the peak of its spectrum at about 250 nm, in the ultraviolet wavelength range. Given that Sirius has a higher T_{eff} than the Sun, its blackbody spectrum is shifted towards shorter wavelengths, hence it is much brighter than the Sun in the U filter compared to the V filter.

As mentioned in chapter 3 the blackbody spectrum that emerges from the photosphere is more or less slightly modified by the formation of spectral lines when the photons cross the surrounding cooler and less dense layers of the stellar atmosphere. This is shown in Fig. 6.1,

---

[6]Not surprisingly, the human eye is most sensitive to light with approximately this wavelength.

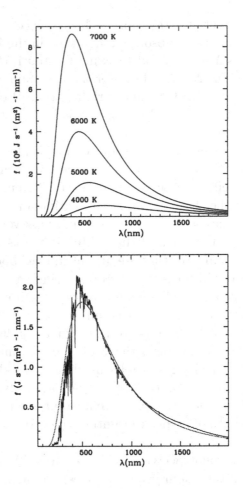

Figure 6.1 *Upper panel:* Brightness of blackbodies of different temperature as a function of the wavelength λ. *Lower panel:* As the upper panel, but for the spectrum of the Sun measured above the atmosphere [76], and a blackbody with a temperature of 5772 K, that gives off the same total amount of energy per unit area and unit time as the Sun (dotted line).

that also displays the observed solar spectrum and the blackbody with temperature T equal to the solar T_{eff}.

If absolute magnitudes are a proxy for the luminosity of stars, colours are a proxy for their T_{eff}. In astronomy a colour is simply the difference of apparent magnitudes in two filters. Let's consider, as an example, the commonly used colour $(B - V)$ defined as the difference $(m_B - m_V)$. If we recall the definition of apparent magnitudes with Vega as reference star, and assuming the extinction has been corrected for, then $(B - V)_0 = m_{B,0} - m_{V,0} = -2.5 \log(f_B/f_{B,Vega}) + 2.5 \log(f_V/f_{V,Vega}) = -2.5 \log(f_B/f_V) - 2.5 \log(f_{V,Vega}/f_{B,Vega})$.

First of all, being a difference of magnitudes, the measured colour is the same as if calculated using the difference of the absolute magnitudes, meaning that it is independent of distance. Also, we can rewrite the colour as $(B - V)_0 = -2.5 \log(f_B/f_V) + \text{constant}$ (the constant depends on the ratio $(f_{V,Vega}/f_{B,Vega})$, that has the same value for all stars). Written in this form, it is clear that a colour is just a measure of the ratio of the brightness in two different wavelength ranges. Looking now at Fig. 6.1, it is easy to realise that this ratio changes when the blackbody temperature changes. As an example, we can take the brightness of a blackbody at 400 nm as representative of the B filter, and at 600 nm as representative of the V filter: The figure shows that at 7000 K the brightness in B is larger than in V, and at 5000 K the reverse is true. This means that $(B - V)$ increases for decreasing temperatures.

If we do not correct the apparent magnitudes for the effect of extinction, then the measured colour is equal to $(B - V) = m_{B,0} + A_B - m_{V,0} - A_V = m_{B,0} - m_{V,0} + (0.3A_V)$ (recalling that $A_B = 1.3A_V$). This colour is larger than the intrinsic one because of the positive term $0.3A_V$, and larger $(B - V)$ means a blackbody at lower temperature, with the peak emission shifted to the red (longer wavelengths). For this reason, the effect of extinction on the measured colours of stars is named reddening. The difference between the measured and intrinsic $(B - V)$ colour is denoted with $E(B - V)$ (more in general, the amount reddening that affects a generic colour $(W_1 - W_2)$ is denoted with $E(W_1 - W_2)$). In the case of the colour $(V - I)$ often used in the next chapter, $(V - I) = m_{V,0} - m_{I,0} + (0.4A_V)$, meaning that the effect of extinction is slightly more severe on this colour compared to $(B - V)$[7].

The measured colours of stars can be used to determine the extinction along the line of sight to astronomical targets following this simple idea. First, we need to establish the intrinsic colours – as many

---

[7]In the diagrams that follow I will usually employ the same symbols for both observed and extinction-free magnitudes and colours. I will use for example m_V and $(B - V)$ to denote the observed values, but also the extinction-free ones, as often found in the literature. The figure captions will explain whether the correction for extinction has been applied.

as possible, to cover a wide wavelength range – of stars in a chosen evolutionary phase. This can be done for example using stars within a distance of around 150–200 light-years; observations of the interstellar medium show that the extinction is negligible within this range. We can then consider the observed colours of samples of more distant stars in the same evolutionary phase (and the same metallicities) of this template zero-extinction sample, distributed along different lines of sight: Comparisons with the local template, give us the values of reddening in each individual colour, which will differ from each other because of the different filters involved. Using the extinction law we can then translate these reddenings to values of the extinction A_V, or the extinction in any other filter.

To compare theory with the extinction-free measurements of these proxies for stellar luminosities and effective temperatures, we need to translate L and T_{eff} of stellar tracks and isochrones to absolute magnitudes and colours in the chosen photometric filters. This step requires the calculations of what we call bolometric corrections (denoted with BC_λ). It is a bit of a tedious affair, that can be summarized as follows.

Each point along an evolutionary track or isochrone must be associated with a spectrum, which can be computed using calculations of stellar model atmospheres. For precision analyses, we cannot use the simple formulas of a blackbody with the model T_{eff} because despite the close similarity, the presence of spectral lines needs to be taken into account for accurate comparisons. Model atmosphere calculations determine the structure of the low-density layers that surround the photosphere (and contain a negligible amount of mass compared to the total mass of a star) and the spectrum to assign to the underlying stellar model. They are performed with computer codes that solve different sets of equations compared to the equations of stellar structure and evolution: In particular, the equation describing the radiative energy transport is different and much more complicated. The input parameters of a model atmosphere are the chemical composition, the same as the composition of the photosphere of the underlying stellar model, the photospheric T_{eff} and surface gravity g, proportional to the ratio between the mass and the square of the radius of the photosphere. Once we have the spectrum associated to a point along an evolutionary track or an isochrone, we calculate the absolute magnitude of that point, for example in the filter V, as $M_V = -2.5\log(L/L_\odot) - BC_V$ where $L_\odot$ is the solar luminosity and BC_V is a constant, named bolometric correction. Its value depends on the spectrum, hence on T_{eff}, g, and chemical composition; it accounts for the portion of the spectrum within the wavelength range of the photometric filter, the brightness of Vega in that filter, and the effect of a distance of 10 pc. After bolometric corrections are computed for the relevant filters, model colours are readily calculated as differences of absolute magnitudes. For example, in the case of the $(V - I)$ colour, $(V - I) = (M_V - M_I) = -2.5\log(L/L_\odot) - BC_V + 2.5\log(L/L_\odot) + BC_I = BC_I - BC_V$. Hence theoretical colours are simply differences of bolometric corrections, which are sensitive to the model T_{eff}, for a fixed chemical composition and g.

Finally, using observed magnitudes and colours as proxies for L and T_{eff} we can produce a 'colour magnitude diagram' (CMD), whereby the horizontal axis displays a star (or model) colour, while the vertical axis displays the corresponding magnitude[8]. The CMDs are just a different way to represent HRDs like those by Russell and Rosenberg shown in chapter 3, without using spectroscopy. They are the workhorse of stellar astrophysics and in particular stellar age determinations, as I will show in the next chapter: I must have drawn on blackboards and whiteboards thousands of CMDs in my lectures on stellar evolution and stellar populations.

6.2 CHEMICAL COMPOSITION

The detailed structure and evolution of stellar models of a given mass depend on the initial chemical composition of the calculations. To use models and isochrones for the determinations of stellar ages, we need to link the abundances entering our calculations with what is measured by stellar spectroscopy.

As mentioned in chapter 4, the initial chemical composition of a stellar model is given in terms of the hydrogen mass fraction X, the helium mass fraction Y, and the metallicity Z. When we input a value for Z in our calculations, we need to assume a set of relative abundances among all heavy elements, because individual abundances are critical to calculating properly the opacities of the stellar matter, the nuclear energy generation, and to be able to keep track of the detailed evolution of the various elements. Observations tell us that for the Sun and most stars in the disk of the Milky Way, about 97% of the initial total mass fraction of metals is contained in a few elements, namely C, N, O, Ne,

---

[8]The first CMDs were published in 1927 by Paul ten Bruggencate in his book *Sternhaufen* (*Star clusters* in German), using observations of globular clusters and open clusters.

Mg, silicon (Si), S, and Fe. Just oxygen makes about 50% of all metals, while Fe, for example, makes about 7% of the initial Z. When moving to stars in the halo (and the bulge) of our galaxy, or stars in the disk below a certain metallicity, the relative distribution of the abundances of metals at fixed Z changes, with O, Ne, Mg, Si, and S enhanced, and C, N, Fe depleted compared to the metal distribution of disk stars. In this case, oxygen makes about 60–70% of the total metallicity, whilst Fe contributes only by about 3%. As a consequence, if we fix Z, a model calculated with the heavy element distribution of stars in the disk will have less oxygen and more iron than a model calculated with the same Z but the metal distribution of halo stars.

The analysis of the observed spectra of stars provides us with T_{eff}, g, and chemical abundances of the targets. Early spectroscopic measurements determined abundance ratios of chemical elements to the solar values because differential analyses minimize the effect of uncertainties in the physics entering the calculations of line formation. Even when absolute abundances (not relative to a reference star) are measured today, it is often still customary to express them relative to the Sun. Iron, in particular, is readily measured because of the large number of spectral lines at optical wavelengths, and it is common to use as a proxy for Z the quantity $[Fe/H] = \log(n_{Fe}/n_H)_{star} - \log(n_{Fe}/n_H)_\odot$, where n_{Fe} and n_H denote, respectively, the number fraction of iron and hydrogen atoms in the atmosphere of the target star and in the Sun[9]. A value $[Fe/H]=-1$ means for example that the measured ratio (n_{Fe}/n_H) is 10 times lower than in the Sun. The spectroscopic measurements of $[Fe/H]$ can be easily compared with photospheric abundances provided by stellar models, because $[Fe/H] = \log(n_{Fe}/n_H)_{star} - \log(n_{Fe}/n_H)_\odot = \log(X_{Fe}/X)_{model} - \log(X_{Fe}/X)_\odot$ where X_{Fe} here denotes the mass fraction of Fe and X the hydrogen mass fraction.

If the effect of the mixing processes included in the term f_{mix} entering the equations of stellar structure and evolution (see chapter 4) is negligible for the surface abundances of the target star, then the observed $[Fe/H]$ is the same as the initial value; given that $(X_{Fe}/X)_\odot$ is known, from the measured $[Fe/H]$ we can determine the initial (X_{Fe}/X) for the model. Measurements of the abundance ratio of additional elements like O and Mg to Fe tell us what are the relative initial abundances of the various metals at fixed Z (see the beginning of this

---

[9]The number fraction of an element is the ratio of the number of its atoms to the total number of atoms in the gas.

section) so that we can finally derive the initial value of Z/X. This is the case, for example, of RGB stars, because their deep convective envelopes erase the effect of the other element transport mechanisms described by f_{mix}[10].

If the models predict that the measured surface abundances of the target star have been affected by the mixings described by f_{mix}, an iterative procedure is necessary. We fix the initial Z/X of the models, assuming that the observed [Fe/H] is equal to the initial value, and see by how much $(X_{Fe}/X)_\odot$ is expected to change for the target star. We then add to the observed [Fe/H] the difference from the initial value predicted by the stellar evolution calculations and rederive the initial Z/X. Another calculation with these new initial abundances follows: If the observed [Fe/H] is not matched we adjust the initial Z/X until the measured [Fe/H] is reproduced after one or more iterations.

To identify the unique set of stellar model abundances that match the abundances of a target star, we then need to know the amount of helium at the star photosphere. The ratio of metal abundances to hydrogen does not fix their absolute values if we do not know also the mass fraction of helium[11]. Unfortunately, the helium content cannot be properly measured in stars with T_{eff} below 8,000–10,000 K, because they do not show helium absorption lines in their spectra[12].

What we do in practice is to determine an average relationship between Y and Z by considering the Big-Bang helium abundance $Y=0.25$ for $Z=0$ (essentially no metals produced at the Big-Bang) and for example the initial helium abundance in the Sun at the initial solar metallicity. From these values we calculate the ratio $\Delta Y/\Delta Z$ that tells us by how much on average Y has increased compared to the Big-Bang value (ΔY), when the metallicity increases from zero (at the Big-Bang) to a given Z (ΔZ). Another possibility is for example to measure the helium abundance in a sample of H_{II} regions, gas clouds around young massive and hot stars, that are associated to regions of star formation

---

[10]Thanks mainly to spectroscopical analyses of RGB stars we have learned that the initial abundance ratios of the various metals are not the same for all stars, as described before.

[11]If we know Z/X and Y, then we can use the fact that $X+Y+Z=1$, to obtain the individual values of X and Z.

[12]Helium emission lines appear in spectra of these 'cold' stars, but they are produced in the upper layers of the stellar atmospheres, the so-called chromosphere, whose theoretical modelling is quite uncertain.

in the disks of spiral galaxies. These hot clouds display emission lines[13] of helium that are used to determine Y, and also of metals like oxygen, to estimate Z. Given the association with star formation, their composition should reflect the actual trend of Y with Z in stars. Observations of several of these regions in the Milky Way and also other galaxies provide an average $\Delta Y/\Delta Z$, typically between 1 and 2, that can be used in model calculations [6].

6.3 DIRECT MEASUREMENTS OF RADIUS, LUMINOSITY AND TEMPERATURE

There are some stars for which we can more or less directly determine radius, T_{eff} and L, but this is possible only in a limited number of cases.

If we can resolve the disk of bright, nearby stars of known distance, measurements of their angular size – the amount of space an object takes up in the field of view of an observer, measured in degrees or fractions of degree[14] – and of the total brightness over the entire wavelength range of their light provide L, T_{eff}, and R[15]. These measurements are very 'expensive' in terms of resources, because there are no single detectors sensitive to photons across the whole electromagnetic spectrum. As you can easily imagine, this method can be applied only to small samples of bright stars relatively close.

For stars in eclipsing binary systems, we can determine distance, mass, radius, T_{eff} – hence the luminosities of their components from the measured effective temperature and radius – as described in the

---

[13]Emission spectra are produced by low density, hot gases in which atoms do not experience many collisions. The lines are caused by photons of discrete energies emitted when electrons in excited atomic states in the gas make transitions back to lower energy levels. An example of astronomical objects with emission line spectra are these clouds made mainly of hydrogen, that orbit around spiral galaxy disks. Their spectra display emission lines when they are heated up by nearby young, massive, and hot stars.

[14]The full Moon, for example, has typically an angular size of about 31 arcminutes, where 1 arcminute (made of 60 arcseconds) is equal to 1/60 of a degree. The angular size depends on both diameter and distance of the target; the Sun has just about the same angular size of the full Moon and this is why we have full solar eclipses, but the physical diameter of the Sun (what we would measure with a ruler if we were on the surface of the Sun) is about 400 times larger than the diameter of the Moon, while the Moon is about 400 times closer to Earth.

[15]The angular radius gives R, and the measured brightness gives L when the distance is known. From this, we obtain T_{eff} using $L = 4\pi R^2 \sigma T_{eff}^4$ (see chapter 4).

appendix. Apart from the fact that only a small fraction of stars is actually in eclipsing binaries, spectroscopy – together with photometry – is needed to obtain L, T_{eff}, and R for the binary components. But spectroscopy cannot reach objects as faint as when we just measure magnitudes in a given photometric filter. The reason is simply related to photon numbers. To detect individual spectral lines we need to measure photons within wavelength ranges on the order of 0.01–0.001 nanometres, while measurements of magnitudes in photometric filters like V or I take advantage of all photons within a 100–200 nm wavelength range.

The same is true for the determination of chemical abundances and T_{eff} (and surface gravity) of single stars from spectroscopy. As an example, today we cannot generally measure the chemical composition of stars on the main sequence of Milky Way globular clusters, while we can measure magnitudes and colours of stars much fainter than the main sequence turn-off.

Another way to determine the mass and radius of single stars is through asteroseismology (see chapter 5). From accurate asteroseismic observations we can measure two quantities, called ν_{max} and $\Delta\nu$, both related to the frequencies of the non-radial pulsations (it is not important to go into the details of how these quantities are measured). When compared to the corresponding values measured for the Sun, ν_{max} and $\Delta\nu$ can provide us with the mass and the radius of the target star if its T_{eff} is known for example from spectroscopy [21]. These measurements require the detection of tiny periodic variations of brightness that are possible only for relatively nearby and bright stars.

GLOSSARY

Bolometric correction: A quantity (that depends on the effective temperature, chemical composition and surface gravity of the model stellar atmosphere) required to compute the magnitude in a given photometric filter, once the total energy emitted per unit time is known.

Colour: Difference of the magnitudes in two different photometric filters.

Colour-Magnitude diagram: Diagram that displays on the horizontal axis a colour of a star (or a stellar model), and on the vertical axis the corresponding magnitude in a given photometric filter.

Magnitude: Measure of the brightness of a star relative to a reference object.

Stellar model atmosphere: Tabulation of the predicted values of physical quantities like temperature, pressure, density, from the base to the top of the atmosphere of a star with a given effective temperature, surface (photospheric) gravity and chemical composition.

FURTHER READING

M. S. Bessell. Standard Photometric Systems. *Annual Review of Astronomy and Astrophysics*, 43:293, 2005

F. R. Chromey. *To Measure the Sky*. Cambridge University Press, 2016

CHAPTER 7

This Is What We Do

It's been a long journey, but the stage is finally set to discuss the most direct methods we routinely employ to determine the age of stars. There is an array of techniques we can deploy, all based on relatively simple ideas built on our knowledge of physics and stellar evolution. The method of choice depends on the observations available, that are typically constrained by the distance and brightness of the target. I am going to start with the Sun, whose age, perhaps surprisingly, is actually not determined using stellar evolution theory.

7.1 THE SUN

As discussed in chapter 5, theoretical models for the Sun (solar models) have been – and are – crucial to test stellar evolution theory and the physics inputs of stellar evolution calculations. We know with very high precision the solar mass ($M_\odot$), luminosity ($L_\odot$) and radius ($R_\odot$ – hence $T_{eff,\odot}$)[1], plus the actual value of the metal-to-hydrogen mass abundance ratio $(Z/X)_\odot$[2], because all relevant heavy elements can be measured for the Sun. We also have measurements of the neutrino flux and the frequencies of a large number of modes of non-radial pulsations, both diagnostics of the internal structure of the star.

The solar mass (about 333,000 times the mass of the Earth) is accurate to about 0.004%, the radius of the photosphere (about 109 times

---

[1]They are: $M_\odot$=1.988 × 10^{30} Kg, $L_\odot$=3.828 × 10^{26} J/s (Joules per second), $R_\odot$=6.957 × 10^8 m, and $T_{eff,\odot}$=5772 K.

[2]There is some debate about the exact value of this ratio. Independent measurements give values for $(Z/X)_\odot$ between 0.018 and 0.025. Models computed with the larger value lead to a better agreement with the sound speed inferred from the frequencies of non-radial pulsations (see also chapter 5).

the radius of the Earth) to better than 0.1%, while the accuracy of the luminosity is limited by its total variation of around 0.1% during the 11-year solar cycle (the adopted value of $L_\odot$ is the average over a solar cycle)[3]. Observations also tell us that the Sun is losing mass (through the solar wind, whose perhaps more spectacular manifestations are the 'northern lights') with a very low rate of just $10^{-13} M_\odot$ per year, and that it spins around its axis very slowly [4]. As a consequence, we can safely neglect mass loss and also rotation when calculating models for the Sun.

To determine its age, we could ideally compute $1 M_\odot$ stellar models considering only convection and gravitational settling in the term f_{mix} of the equations of stellar structure (see chapter 4), and choose a pair of initial Y and Z abundances such that, when the model matches the observed radius and luminosity of the Sun[5], the actual measured ratio $(Z/X)_\odot$ is also matched[6]. Getting the correct initial chemical composition is crucial because it affects lifetime, luminosities, and radii of the models.

We may need various trial-and-error calculations to match all three quantities, because the luminosity and radius of the Sun are determined both by the unknown age and initial chemical composition. Age, in turn, also determines how much the surface Y and Z decrease from the initial values, due to gravitational settling. When the match is achieved, the age of the corresponding model gives us the solar age, and the model inner structure can be compared with asteroseismic results and constraints from the solar neutrino flux.

This is however not a viable procedure, because we can find different combinations of initial Z and Y that can eventually lead to match

---

[3]The solar cycle is a nearly periodic 11-year change of the 'activity' of the Sun. 'Stellar activity' is a phenomenon typical of MS stars with convective envelopes whose more characteristic manifestation is the appearance of starspots, or sunspots in case of the Sun. These are spots darker than the surrounding areas on the surface of the star, caused by magnetic fields. Stellar activity is usually not included in stellar evolution models, but its effect on the internal structure and lifetimes of the models is negligible.

[4]The period of rotation is about 25 days at the equator, and about 32 days at a latitude of 60°. The speed of rotation is different at different latitudes because the Sun is not a solid body like Earth.

[5]When radius and luminosity are matched, also T_{eff} is matched, because these three quantities are related, as described in chapter 4.

[6]Gravitational settling tends to change with time the surface abundance of elements in MS models, decreasing Z and Y while increasing X.

$L_\odot$ and $(Z/X)_\odot$ for different ages of the model[7]. An additional problem is that there is a free parameter, a number we have to choose at the beginning of stellar model calculations, in addition to mass, X, Y and Z. This parameter is called 'mixing length', and it is important when there is convection. If we think of convection in terms of what we see in a pot of boiling water, this number tells us how far the gas bubbles travel before dissolving and releasing their heat in the surroundings. We cannot fix the value of the mixing length from first principles, and laboratory experiments cannot reproduce the conditions of the convective regions in stars: We just know that its value should be between 1 and 2, or slightly more. Different values of the mixing length change the evolution of the radius of models with convective outer layers, while their internal structure, the energy generation, luminosity, and lifetimes are unaffected, even when the inner regions are convective. It only matters when there is convection in the outer layers, and just for the radius (hence also T_{eff}) of the models.

Given the high accuracy of the solar radius and the fact that solar models predict a convective envelope – plus we see convective cells on the photosphere of the Sun – it is natural to use the Sun to fix this number. But to determine unambiguously the mixing length, plus the solar initial Y and Z, we need to know its age. If we know the age of the Sun, then we can adjust the mixing length and initial Y and Z of our $1 M_\odot$ stellar evolution calculations, to match simultaneously $R_\odot$, $L_\odot$, and $(Z/X)_\odot$ when the models reach the solar age. In this way, there is a unique solution for these parameters, no ambiguity. This is how we compute solar models to be tested against helioseismology and neutrino measurements.

After reading this, we might ask ourselves whether the agreement of solar models with helioseismology and neutrino flux is just the consequence of a circular argument: It seems like we are forcing our models

---

[7]An increase of the initial Z, can be compensated by a suitable decrease of Y that in turn might increase X – let's recall that $X+Y+Z=1$. The solar helium cannot be measured because the T_{eff} of the Sun is too low to have He absorption lines in the spectrum. There are emission lines, but they are difficult to interpret in terms of helium abundance. Perhaps ironically, helium was discovered observing the outer edge of the Sun during the solar eclipse of 1868 by Jules Janssen and Norman Lockyer, who detected one emission line at optical wavelengths that could not be associated with any known element. Hence the name helium, derived from 'helios', the Greek name for the Sun. Helium remained undiscovered on Earth until Luigi Palmieri detected the same spectral line in 1882, during the analysis of the lava of Mount Vesuvius, in Italy.

to be a perfect match to the Sun. Actually, this is not the case. We are just finding the best possible estimate of the initial chemical composition for our model, a crucial prerequisite to testing the internal structure of models against helioseismology and neutrino observations. The mixing length does not affect the internal structure of the models, and its calibration on the Sun enables us to fix this undetermined quantity in all stellar evolution calculations. If the physics in our models is wrong or highly inaccurate, this 'calibration' won't be able to hide the fact that our solar model won't match results from helioseismology and neutrino observations.

The question now is: How do we determine the age of the Sun? The best estimate of the age of the Sun actually comes from measurements of the age of meteorites, using techniques from geology. Meteorites are fragments of asteroids that in turn are the leftovers of the formation of the Solar System, and their age tells us the age of not only the Solar System but also of the Sun itself. When stars form in giant molecular clouds (see chapter 4) the collapsing clumps are surrounded by a flat, thin disk from which they accrete gas for up to a few million years: Planets form in these disks, and this explains why the orbits of the planets in the Solar System lie more or less all in the same plane – apart from the dwarf planets Pluto and Eris – called 'ecliptic'. This scenario has been confirmed by recent observations of the star PDS 70b (a young star with a mass about 80% of the mass of the Sun) in the constellation Centaurus, at a distance of about 370 light-years, which show a clear image of a disk surrounding the central star and two planets –both more massive than Jupiter– breaking through it.

The age of meteorites – hence of the Sun – has been established very accurately using radioactive elements. As discussed in the appendix, there are 92 chemical elements which occur in nature, and each chemical element is defined by the number of protons in the nucleus. For example, hydrogen has one proton, helium two protons, lithium three protons, and so on. However, there are different 'versions' of the same element – like different styles of trousers – called 'isotopes', which differ by the number of neutrons in the nucleus. Hydrogen (H) atoms are usually made of one proton and one electron, with no neutrons, but there are also the isotopes named deuterium (^{2}H) and tritium (^{3}H), with one proton together with one and two neutrons in the nucleus, respectively. When isotopes are involved in the discussions that follow, the number attached to the symbol of the chemical element denotes the number of particles in the nucleus, the sum of protons and neutrons.

Often isotopes with a number of neutrons much larger than the number of protons are not stable. The neutrons in the nuclei of these elements gradually change into protons usually by emitting an electron (and an antineutrino) in the so-called β decay process – but there are more complex modes of decay like for uranium isotopes. As a result, the unstable isotope transforms into a different element, because the number of protons in the nucleus has changed. The decaying nucleus is usually called parent nucleus (P), the stable nucleus produced by the decay is called daughter nucleus (D), and the total number of particles P+D is constant with time. A material containing decaying nuclei is called 'radioactive'.

Even though the physics behind this process is complex, it is very simple to empirically model the decay of an unstable isotope. Given a sample of unstable nuclei of an element, experiments show us that the probability the decay of one nucleus will happen within a given time interval is a constant – the 'decay constant', which varies between different types of unstable nuclei – the same for all particles in the sample at any time, and independent of the number of nuclei present. This means that if we have for example 1000 nuclei of an unstable isotope with a decay constant of 0.1 per year (10% probability that each nucleus will decay in a time interval of one year), 100 nuclei will have decayed after one year (10% of 1000 nuclei), then 90 nuclei will decay after another year (10% of the 900 nuclei left over after one year), followed by 81 decays in the following year (10% of the 810 nuclei left over after two years), and so on. In a nutshell, the number of unstable nuclei decaying in a given time interval is proportional to the number of unstable nuclei in the sample. In mathematical terms, this relationship is written as $dN/dt = -\lambda N$, where λ is the decay constant and N the number of unstable nuclei at a generic time t[8]. The solution is $N = N_0 e^{-\lambda t}$, where N_0 is the initial value of N: Because the exponent in $e^{-\lambda t}$ is negative, N decreases when the time t increases[9]. Given this simple mathematical form of the solution, it is easy to calculate the so-called half-life (denoted as $\tau_{1/2}$) of an unstable nucleus, defined as the time taken for half the unstable nuclei in a sample to decay, which is equal to $0.693/\lambda$. Values of $\tau_{1/2}$ are then used to compare the timescales for the decay of unstable isotopes. Below I list a few parent

---

[8]dN denotes the variation of the number of nuclei after a time interval dt. The minus sign on the right-hand side of the equation tells us that N has to decrease when the time increases.

[9]The Euler's number 'e' is equal to 2.71828.

unstable nuclei and the daughter stable product of the decay, with the corresponding half-life:

^{14}C $\to$ ^{14}N, $\tau_{1/2}$=5730 yr;

^{87}Rb $\to$ ^{87}Sr, $\tau_{1/2}$=48.8 Gyr;

^{232}Th $\to$ ^{208}Pb, $\tau_{1/2}$=14.05 Gyr;

^{235}U $\to$ ^{207}Pb, $\tau_{1/2}$=0.7 Gyr;

^{238}U $\to$ ^{206}Pb, $\tau_{1/2}$=4.47 Gyr;

where Rb, Th, U, Sr, Pb denote rubidium, thorium, uranium, strontium, and lead, respectively.

In 1946 the chemist Willard Libby realized that ^{14}C can be used as a clock to determine the age of organic matter, and was awarded in 1960 the Nobel Prize for his discovery. Fresh ^{14}C is continuously created by cosmic rays (see footnote 6 in chapter 5) which produce neutrons when interacting with the Earth atmosphere; these neutrons can be captured by stable ^{14}N nuclei, with the emission of a proton. This way, by swapping a proton with a neutron, nitrogen nuclei transform into ^{14}C (carbon nuclei have 6 protons, whilst nitrogen have seven). This ^{14}C, together with the stable and infinitely more abundant ^{12}C (that makes about 99% of carbon on Earth) and ^{13}C (that makes the remaining 1%, while ^{14}C is about 10^{-10}% of the total content of carbon on Earth, a very small but measurable level) combine chemically with oxygen to form carbon monoxide and carbon dioxide[10] which is incorporated into plants by photosynthesis, and passed through the food chain to animals. Every plant and animal will therefore have the same ratio ^{14}C/^{12}C as in the atmosphere, because the decay rate of ^{14}C in a living organism is slow relative to the movement of carbon through the food chain.

This continuous replenishment of ^{14}C from the atmosphere stops abruptly when the organism dies. From that moment on ^{14}C decay takes over, so that after about 5700 years (the half-life of ^{14}C) the amount of ^{14}C relative to ^{12}C (that is stable) in a dead organism is half the amount found in living beings. After about 55,000–70,000 years there is no longer any measurable amount of ^{14}C left. To summarize, a comparison of the ^{14}C/^{12}C ratio measured in a dead organism with the atmospheric ratio can be used as a clock to determine the time of death of the specimen – up to about 55,000–70,000 years ago – because

---

[10]All three isotopes of carbon combine with oxygen in the same way, to make these molecules.

we can predict quantitatively how many ^{14}C nuclei decay to nitrogen in a given time interval.

The same idea can be applied to study the age of rocks and meteorites but using radioactive elements with a longer half-life. One example is ^{87}Rb that decays to ^{87}Sr, with a half-life of almost 50 Gyr. Comparisons of the current ratio of the abundance of the two elements with the initial one can tell us the age of the rock, because ^{87}Rb steadily transforms into the stable ^{87}Sr isotope. But contrary to the case of carbon dating of dead organisms, we do not know the initial abundance ratio of the two elements in our rocks. This problem is however not insurmountable thanks to the following technique, based on measurements of the abundance of ^{87}Rb, ^{87}Sr, and the stable strontium isotope ^{86}Sr, which is not produced by nuclear decays.

Let's have a look at panel (a) of Fig. 7.1, that shows a graph with the (arbitrary) number of ^{87}Rb (horizontal axis) and ^{87}Sr (vertical axis) isotopes in three different minerals of a fictitious rock (or meteorite)[11] at the time of their formation (set to $t=0$). When we measure the chemical composition of the rock at a later time $t=1$, some of the ^{87}Rb nuclei have transformed to ^{87}Sr following the law of radioactive decay (remember that the number of decays in a given time interval depends on the number of unstable nuclei in the sample), and the new abundances are shown in panel (b) of the same figure. The ratios ^{87}Sr/^{87}Rb measured at time $t=1$ are different among the three samples, and we cannot assess whether they have different ages, or they were formed with different initial abundances of ^{87}Sr and ^{87}Rb.

This problem can be solved if we measure the number ratios ^{87}Rb/^{86}Sr and ^{87}Sr/^{86}Sr, as shown in panels (c) and (d). The ratio ^{87}Sr/^{86}Sr does not depend on the abundance of ^{87}Sr when a mineral forms, because chemical processes in nature do not alter the ratio of isotopes of the same element with such a small difference of the number of particles in the nucleus (86 and 87 in this case).

Let's assume the initial number ratio ^{87}Sr/^{86}Sr at formation was equal to 2 (an arbitrary number) for the three minerals; if we use this ratio and the initial numbers of ^{87}Sr and ^{87}Rb nuclei of panel (a), we obtain the diagram in panel (c), a straight line parallel to the horizontal axis. The ^{87}Sr/^{86}Sr is constant, but the ^{87}Rb/^{86}Sr ratio varies among

---

[11] A rock is commonly made of one or more minerals, and a mineral is a compound having a well ordered internal structure, and characteristic chemical composition (quartz is an example of mineral).

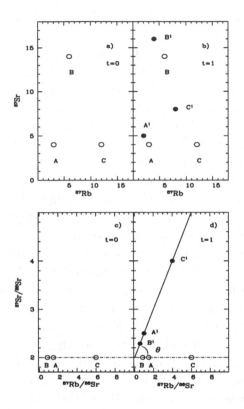

Figure 7.1 Visual explanation of the isochron method to determine ages of rocks and meteorites, using strontium (Sr) and rubidium (Rb) abundances. ^{87}Rb is the parent nucleus, ^{87}Sr the daughter nucleus and ^{86}Sr is a stable isotope of Sr that is not produced by radioactive decays. We consider three minerals in a fictitious rock or meteorite (A, B, C). *Panel (a)*: Number of ^{87}Rb and ^{87}Sr atoms when the rock forms (time $t=0$). *Panel (b)*: Number of ^{87}Rb and ^{87}Sr atoms at a later time $t=1$ (filled circles) compared to the values at $t=0$. The changes are due to the decay of ^{87}Rb to ^{87}Sr. *Panel (c)*: Number of ^{87}Rb and ^{87}Sr atoms when the rocks form (time $t=0$), normalized to the number of the stable ^{86}Sr isotope. *Panel (d)*: Number of ^{87}Rb and ^{87}Sr atoms at $t=1$ normalized to the number of the stable ^{86}Sr isotopes (filled circles – the ratios at $t=0$ are also shown as comparison). These ratios lie on a straight line (an isochron), if the minerals formed at the same time from the same material. The angle (θ) between this line and the horizontal axis depends on the age of the rocks. The higher the age, the larger the angle.

the three samples, because of the different initial numbers of ^{87}Rb nuclei.

When we measure the composition of the samples at time $t=1$, some of the initial ^{87}Rb nuclei have transformed to ^{87}Sr, whilst the number of stable ^{86}Sr isotopes has not changed: This implies that ^{87}Sr/^{86}Sr has increased, and ^{87}Rb/^{86}Sr has decreased, the variation of these two ratios being related by the decay constant of ^{87}Rb. If we now use the values of panel (b) of the figure for the numbers of ^{87}Sr and ^{87}Rb nuclei at $t=1$, we find that also at this time the points corresponding to the three samples lie on a straight line in the ^{87}Rb/^{86}Sr versus ^{87}Sr/^{86}Sr diagram, but the line is at an angle θ compared to the horizontal axis, as shown in panel (d). This angle gets wider with increasing t, because ^{87}Rb/^{86}Sr continues to decrease and ^{87}Sr/^{86}Sr is steadily increasing.

The straight line is called 'isochron'[12] because if the minerals are coeval the measurements displayed in the diagram will lie along this 'equal age' line. If we extend the isochron at time $t=1$ towards lower values of ^{87}Rb/^{86}Sr, we find that for ^{87}Rb/^{86}Sr$=0$ – meaning that no ^{87}Sr nucleus could have been produced by the decay of ^{87}Rb – the initial ratio of Sr isotopes ^{87}Sr/^{86}Sr$=2$ is recovered. Our measurements therefore can tell us both the age of the object and the initial value of the ^{87}Sr/^{86}Sr number ratio[13].

This isochron technique has features that tell us whether or not we can get a reliable age for our rock or meteorite. The underlying assumption is that the samples formed at the same time, from a common pool of material, and the abundances have been altered in time only by radioactive decay. If this is not the case, our measurements of ^{87}Rb/^{86}Sr and ^{87}Sr/^{86}Sr will be scattered in the diagrams of panels (c) and (d), and won't lie close to an isochron. This tells us that we cannot get a meaningful age for the formation of the object.

Another commonly used isochron method to determine the age of the Solar System is based on the uranium ^{235}U and ^{238}U unstable isotopes, which decay to ^{207}Pb and ^{206}Pb, respectively, with different

---

[12]Not to be confused with 'isochrone', which refers to stellar evolution calculations even though it is the same word, with just a slightly different spelling.

[13]These results can be demonstrated to be true for any time t using the equation of radioactive decay for ^{87}Rb. I give here the general equation of this isochron, valid for any time t: $(^{87}\text{Sr}/^{86}\text{Sr})_t = (^{87}\text{Sr}/^{86}\text{Sr})_0 + (^{87}\text{Rb}/^{86}\text{Sr})_t \; (e^{\lambda t-1})$, where the suffix 0 means the number ratio at formation, the suffix t means the value at time t, and λ is the decay constant of ^{87}Rb. This is the equation of a straight line, the angle θ with the horizontal axis given by the relation $(e^{\lambda t-1}) = \tan(\theta)$.

half-lives hence at different speeds. Their use for age-dating was developed independently by Erich Gerling, Arthur Holmes, and Friedrich Houtermans in the 1940s. The idea here is to measure, in addition to the present-day $^{235}U/^{238}U$ number ratio, also the $^{207}Pb/^{204}Pb$ and $^{206}Pb/^{204}Pb$ ratios, where ^{204}Pb is a stable isotope of lead, not produced by decays. It can be shown from the equation of radioactive decay that in a diagram with $^{207}Pb/^{204}Pb$ on the vertical axis and $^{206}Pb/^{204}Pb$ on the horizontal one, an isochron has a slope that depends on the unknown age, the measured value of $^{235}U/^{238}U$, and the decay constant of these two isotopes[14]. An example is shown in Fig. 7.2, which displays two isochrons for 4.5 and 4.6 Gyr respectively, together with measurements from five meteorites, whose common age turns out to be around 4.55 Gyr.

The age of the Sun determined from meteorites and used in stellar evolution is equal to 4.57 Gyr with an error of 20 million years, or less, around this value [5, 25]. The calculation of the solar model then works this way: We start the computation from the PMS with initial trial values of Y, Z, and mixing length, and after about 40 million years the model enters the MS, with radius and luminosities slowly increasing with time. When the model has reached an age of 4.57 Gyr (an error of 20 million years around this age has a negligible effect on the model, because the evolution is very slow on the MS) we check whether it matches the measured $R_\odot$, $L_\odot$, and $(Z/X)_\odot$. If it doesn't, we need to repeat the calculations with different trial values until these three quantities are matched for an age of 4.57 Gyr: The resulting solar model can then be compared with results from helioseismology and neutrino observations. Typically, variations of the mixing length affect essentially only the radius of the models, while changes of Y and Z affect luminosity, radius and lifetimes[15]. As mentioned in chapter 4, initial values for Y and Z of the solar model are about $Y=0.27$, and $Z=0.02$, and the mixing length has a numerical value around 1.8–2.0.

To close this section, let's make an experiment by treating the Sun like any other star we try to give an age to. I can fix Y for a given Z

---

[14]To be more specific, the slope θ of the isochron is $\tan(\theta)=(^{235}U/^{238}U)_t$ $[(e^{\lambda_{235}t}-1)/(e^{\lambda_{238}t}-1)]$, where $(^{235}U/^{238}U)_t$ is the present number ratio of these uranium isotopes, equal to 137.88, t is the age of the isochron, λ_{235} and λ_{238} the decay constants of the uranium isotopes.

[15]As a general guideline, an increase of the mixing length decreases the radius of the model, whilst an increase of the initial Z or a lower Y decrease the model luminosity at the solar age.

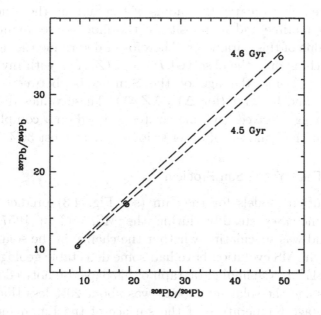

Figure 7.2 Age of five meteorites (from Nuevo Laredo in Mexico, Forest City in Iowa, Modoc in Kansas, Henbury in Australia, and Canyon Diablo in Arizona) with the isochron method applied to lead ^{206}Pb and ^{207}Pb daughter nuclei, produced by the decay of the uranium parent nuclei ^{238}U and ^{235}U, respectively. ^{204}Pb is a stable isotope that is not produced by radioactive decays. Open circles display the meteoritic data [85], and the dashed lines correspond to isochrons for ages equal to 4.5 and 4.6 Gyr.

by using a $\Delta Y/\Delta Z$ relationship derived from H_{II} regions, as discussed in the previous chapter, which typically results in values between 1 and 2. This way the initial chemical composition of the models can be constrained using only spectroscopic measurements, without assuming an age for the Sun. I now determine the age of the Sun by calculating 1 $M_\odot$ models with an arbitrary value of the mixing length, let's say 1.5. I then adjust only the initial Z to match the measured $(Z/X)_\odot$, fixing the corresponding initial Y to the value given by the relationship $Y=0.25+(\Delta Y/\Delta Z)Z$, where $Y=0.25$ is the Big Bang value (for $Z=0$). I am going to use in one case $\Delta Y/\Delta Z=1$, and in the other case $\Delta Y/\Delta Z=2$, to cover the whole range of possibilities. The value of the mixing length is not calibrated on any real star, and I don't expect my models to be able to match the radius of the Sun; on the other hand, I know that lifetime and luminosity of the models are unaffected by the exact value of this parameter. Therefore I determine the age of the Sun by matching only the observed $L_\odot$ and $(Z/X)_\odot$ with my models.

This way, I find the age of the Sun to be between 3.2 (for $\Delta Y/\Delta Z=2$) and 5.6 Gyr (for $\Delta Y/\Delta Z=1$). These values differ from the 4.57 Gyr age derived from meteorites – based on a completely independent method and set of observations – by at most 30%[16].

7.1.1 The Faint Young Sun Problem

Stellar evolution models for the Sun (see Fig. 4.3) predict that its luminosity increases steadily during the MS, and in 1957 Martin Schwarzschild was speculating whether the change in the solar brightness during the MS evolution have had some detectable geological consequences [117]. In the final paragraph of his paper he noticed that two billion years ago the solar luminosity was about 20% less than today, and the average temperature of the surface of the Earth must then have been around the freezing point of water, wondering whether such a low average temperature was too cool for the algae known to have lived at that time.

The geological record shows that, as early as 3–4 Gyr ago, liquid water and temperate-warm climates were present on our planet, although the Sun was about 20–25% less luminous than today, implying that with the current atmospheric composition and continents the Earth should have been experiencing a full glaciation. And once ice-covered, the Earth surface would reflect the light from the Sun more efficiently so that even a higher solar luminosity would not be enough to exit the glaciation. This is, in a nutshell, the 'faint young Sun problem'.

An astrophysical solution to this problem envisages a 5% more massive Sun in the past, for a higher mass on the MS implies a higher solar luminosity, hence a warmer Earth. During the last 3–4 Gyr the mass of the Sun would have progressively decreased, but the associated decrease of luminosity would be more than compensated by the brightening due to the natural evolution along the MS. The problem is that the amount of mass lost over the last 4 Gyr considering the current mass-loss rate of the Sun plus the mass converted to energy in

---

[16]For $\Delta Y/\Delta Z=2$ I get an initial Y larger than in the solar calibration, and a lower initial Z. The opposite is true when $\Delta Y/\Delta Z=1$. The initial Y and Z obtained from the solar calibration lead to $\Delta Y/\Delta Z$ of about 1.3.

the core, amount to just 0.05% of the current solar mass. Much higher mass-loss rates in the past are also most likely ruled out by observations of younger stars with mass and composition similar to the solar values.

Another avenue to find a solution is to look at the Earth's distant past, because the calculations of the effect of a fainter Sun are made in the assumption that the early Earth atmosphere and continents were the same as today. The most recent studies show indeed that the latest constraints on the chemical composition and pressure of the Earth atmosphere, the fraction of land, and the ocean temperatures at the time, coupled to sophisticated three-dimensional models for the global climate, can solve the problem [22]. Higher (compared to current values) atmospheric concentrations of carbon dioxide and methane could increase more efficiently the atmospheric temperature due to the absorption of radiation from the surface of the planet. This phenomenon, together with a small contribution from a lower fraction of emerged land – that causes a less efficient reflection of sunlight – can explain the mild temperatures of 3–4 Gyr ago, despite a fainter Sun.

7.2 AGE OF INDIVIDUAL STARS

I discuss now the determination of the age of individual stars other than the Sun, describing several techniques that can be used depending on the available data. We can start with the case of the relatively few stars of known mass and radius. They can be the components of eclipsing binary systems, or for example single RGB stars with mass and radius obtained from asteroseismology. Their data can be compared with predictions from theoretical isochrones (the stellar evolution isochrones, not the isochron of radioactive dating) in a mass-radius diagram, as shown in Fig. 7.3.

The left-hand panel shows results – with typical error bars from observations – for three fictitious stars[17] compared with masses and radii of theoretical isochrones with several ages and two different metallicities (labelled by the quantity [Fe/H] discussed in chapter 6.). The two lower mass objects are on the MS, whilst the more massive one is on the RGB.

---

[17]They could be the component of an eclipsing binary, plus a single star studied with asteroseismology, for example.

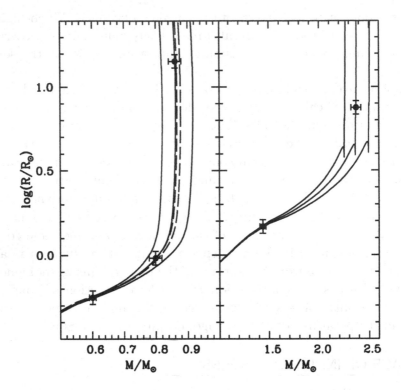

Figure 7.3 *Left Panel:* Mass-radius (in solar units) diagram of three fictitious stars (displayed as filled circles): One is on the MS, one around
the turn-off, and one on the RGB (in order of increasing mass). The
error bars associated to this fictitious data correspond to a 2% error
on the mass and 10% error on the radius. Solid lines display isochrones
for 8, 10 and 12 Gyr (younger isochrones have a higher mass evolving on the RGB) and initial chemical composition with [Fe/H]=−1.3,
Y=0.246. Dashed lines correspond to isochrones with [Fe/H] increased
by a factor of 2 (0.3 dex) and ages equal to 10 and 10.5 Gyr, respectively. *Right Panel:* As the left panel but for two fictitious stars and
isochrones with solar initial chemical composition and ages equal to
600, 700, and 800 Myr, respectively.

Their age is equal to the age of the isochrone that crosses the point
corresponding to the observed M and R. The error on the age is given
by the age range of the isochrones that sweep the area enclosed by the
error bars of the observations. This is a method to derive ages that has
the advantage of not requiring an estimate of the distance to the stars.

For the components of an eclipsing binary system – expected to be formed at the same time with the same initial chemical composition – the derived ages have to be the same within the errors. If this is not the case, it might be that the observations have some error, or there is a problem with the theoretical isochrones.

It is easy to notice that if a star is on the MS and far from the turn-off (the turn-off in this diagram is where the isochrone changes slope to become almost vertical), its position in this diagram is fairly insensitive to age; this is due to the very long MS lifetimes and slow change of radius compared to any other stage of evolution. If such a MS star belongs to an eclipsing binary, its age can be determined from the age the more massive and more evolved companion, if any, as discussed below. If both components are MS stars with mass much lower than the Sun, it is not possible to estimate an even vaguely precise age, because they both hardly change their radius over the age of the universe. Also, given that the post-MS portion of an isochrone is populated by stars with virtually constant mass, a measure of just the mass of the RGB star in the figure effectively fixes its age, irrespective of the exact value of the radius.

Assuming the initial metallicity of these fictitious stars is equal to [Fe/H]−1.3, the best estimate of the age of the RGB and the MS object close to the turn-off is 10 Gyr, with a total range of possible ages – due to the errors on the measurements of M and R – between 8 and 12 Gyr for the turn-off star, and between about 9 and 11 Gyr for the RGB star.

If we use isochrones with a wrong initial metallicity, for example because we lack accurate spectroscopic measurements, our derived age can be wrong. In this example a metallicity too high by a factor of two increases the best age estimate (the age of the isochrone that exactly matches the measured values of M and R) for the turn-off and RGB stars by about 0.5 Gyr, compared to the value obtained with the correct initial [Fe/H].

The right-hand panel of Fig. 7.3 displays the case of the components of a fictitious eclipsing binary with initial solar metallicity, and three isochrones with the appropriate composition. The lower mass component is still on the MS, while the more massive companion is on the RGB. The best estimate of the age of this system is 700 Myr, as derived from the RGB component (the age of the MS star is undetermined), with an error of much less than 100 Myr around this value.

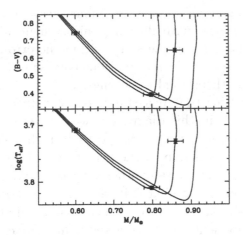

Figure 7.4 Same fictitious stars and isochrones as in the left-hand panel of Fig. 7.3, but in a M-$(B-V)$ (top panel) and a M-$\log(T_{eff})$ (bottom panel) diagram. Error bars on the position of the stars in these diagrams are equal to 0.01 mag in $(B-V)$ and 50 K in T_{eff}, respectively.

The following Fig. 7.4 displays the case of using measurements of mass and a colour $((B-V)$ in this case) or effective temperature, instead of radius. These M-$(B-V)$ and M-$\log(T_{eff})$ diagrams are also independent of the distance to the targets, but the measured $(B-V)$ value needs to be corrected for the effect of reddening when compared to the intrinsic colours provided by the isochrones. The three fictitious stars are the same as in the left-hand panel of Fig. 7.3, and again, the more advanced the evolutionary phase, the more sensitive are these diagrams to age, as discussed for the mass-radius graph. The difference is that using a colour or the effective temperature instead of the radius gives the chance to put some constraints also on the age of the MS star. In fact, even at this low mass, isochrones of different ages show some separation in these diagrams (for this star both effective temperature and colour change more with age than the radius).

What if we cannot measure the mass of our targets, as it is actually the case for the majority of stars? Figure 7.5 shows two 'workhorses' of stellar age determinations, the CMD and the 'Kiel diagram', a graph with the star effective temperature on the horizontal axis and the sur-

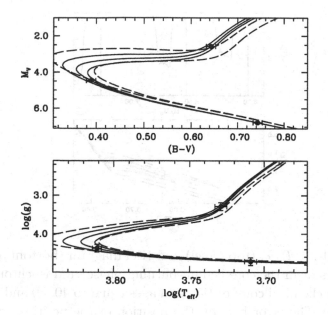

Figure 7.5 The same fictitious stars as in the left panel of Fig. 7.3 but in a CMD (upper panel) and a Kiel diagram (bottom panel). The solid lines display the same isochrones as in the left panel of Fig. 7.3, whilst the dashed lines show a 5 Gyr and a 10 Gyr isochrones, both with [Fe/H] increased by a factor of 2. Error bars on the position of the stars are equal to 50 K in T_{eff}, 0.1 dex in $\log(g)$, 0.01 mag in $(B-V)$, and 0.05 mag in M_V.

face gravity g on the vertical axis[18], both usually determined from spectroscopy.

There is also a photometric system named after Bengt Strömgren, that provides proxies for both g and T_{eff}. It comprises the four filters u, v, b, y (plus other two filters not relevant to this discussion) centred between 350 and 550 nm, with widths of 20–30 nm. A colour like $(b-y)$ is a proxy for T_{eff}, while the index $c_1 = (u - v) - (v - b)$ is a proxy for the surface gravity. A diagram with $(b - y)$ on the horizontal axis and c_1 on the vertical axis is equivalent to a Kiel diagram.

---

[18]In stellar astrophysics the increase of speed of a body for each second of free fall due to gravity is measured in centimetres per second, not in metres per second as usual in physics. In these units the numerical value of g at the Earth surface becomes 981, instead of 9.81.

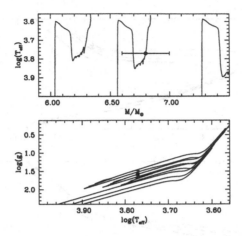

Figure 7.6 M-log(T_{eff}) (top panel) and Kiel diagram (bottom panel) of a fictitious star in the core helium-burning phase, and isochrones with solar initial chemical composition and ages equal to 40, 50 and 60 Myr, respectively. The error bars on the position of the fictitious stars are equal to 50 K in T_{eff}, 0.1 dex in log(g), and 6% in mass.

Figure 7.5 shows that the shape of an isochrone in the Kiel diagram is pretty much the same as in a CMD or HRD, the surface gravity behaving like the luminosity of the star. Just to give an idea of the position of the Sun in these diagrams, $M_{V,\odot}$=4.8, $(B - V)_\odot$=0.65, log($T_{eff,\odot}$)=3.76, and log($g_\odot$)=4.44[19].

Like for the mass-radius diagrams, the age of an individual star is obtained by finding the isochrone of the appropriate initial metallicity, that matches the position of the target star. Age determinations with the CMD require the knowledge of the star distance (for example from parallax measurements) and extinction, to compare with absolute magnitudes and intrinsic colours predicted by the isochrones; the Kiel diagram is instead independent of these two quantities[20].

The most age-sensitive region of these diagrams is the turn-off, where isochrones of different ages are much more separated compared to the RGB and MS. The RGB is weakly sensitive to age in these two diagrams, making the age estimates of RGB stars very uncertain if we

---

[19]For comparison, the surface gravity of Earth gives log(g)=2.99 in these units.
[20]But its $(b - y)$-c_1 proxy is affected by extinction.

do not have information about their mass. The MS below the turn-off is also insensitive to age, as for the mass-radius diagram.

The three fictitious stars in this figure are the same ones as in Fig. 7.3, with typical error bars on their measured values of magnitudes, colours, effective temperatures and surface gravities. The age of the star closer to the turn-off – assuming its initial metallicity corresponds to [Fe/H]=−1.3 – is again between 8 and 12 Gyr, when accounting for the error bars. The age of the MS star cannot be estimated, whilst the age of the RGB star has a much larger indetermination compared to Fig. 7.3.

Measurements of chemical composition are, again, necessary to fix the appropriate initial metallicity of the isochrones. Isochrones with an incorrect metallicity can lead to ages systematically different from the true ones, even by several billion years. Let's imagine again to use a metallicity too high by a factor of two in our age determination; the CMD of Fig. 7.5 shows that, in this case, a 5 Gyr isochrone at the base of the RGB matches the position of the fictitious RGB star. This age is a few billion years below the lowest possible age (compatible with the measurement errors) obtained using isochrones with the correct initial metallicity of the star. The same is true also for the star close to the turn-off region. A similar result is obtained if we use the Kiel diagram instead.

As an example of a younger star, Fig. 7.6 shows a mass-effective temperature and a Kiel diagram of a $6.7M_\odot$ star in the core He-burning phase. The age derived from isochrones in the Kiel-diagram is between about 40 and 60 Myr, when taking into account typical error bars on the observations, while the knowledge of the mass makes the age determination much more precise, as shown by the M-T_{eff} diagram.

7.2.1 Ages from Radioactive Elements

As in the case of rocks and meteorites, unstable nuclei can also be used to determine the age of stars. This is a method completely independent of measurements of mass, radius, T_{eff}, surface gravity, magnitudes, and does not rely on stellar evolution models. In 1929 Ernest Rutherford, the father of nuclear physics who, among others, introduced the concept of half-life of an unstable nucleus, tried to estimate the age of the radioactive uranium found on Earth. He obtained an age of about 4 Gyr for our star, under the (erroneous) assumption that uranium is produced by the Sun. Later on, in 1960 Fowler and Hoyle discussed in detail the use of the long-lived radioactive isotopes ^{232}Th, ^{235}U, and ^{238}U measured in stars, to determine the age of the Milky Way [37].

If we look at the previous short list of radioactive elements and their half-life, we see that ^{232}Th, ^{235}U, and ^{238}U have values of $\tau_{1/2}$ comparable to the ages of even the oldest stars. Thorium (with 90 protons in the nucleus) and uranium (92 protons) – and all elements with a number of protons larger than iron – are also made inside stars, but not by nuclear fusion. As mentioned briefly in chapter 4, nuclear fusion can make elements only up to Fe (26 protons); heavier nuclei are produced by neutron captures onto the lighter nuclei made by fusion. Whenever there is a high production of neutrons through some nuclear reactions, the surrounding nuclei can capture them to produce heavier (with more neutrons) isotopes of the same element. Eventually, these isotopes become unstable due to the too large number of neutrons and – depending on their half-life – decay to a new, stable element with a higher number of protons. Then neutron captures start again on this nucleus, and the cycle repeats until uranium is produced.

These neutron capture processes are named rapid (r-process) or slow (s-process). In the r-process an unstable isotope captures several more neutrons before decaying to a stable element, because of the large flux of neutrons produced by ongoing nuclear reactions, whilst in the s-process the unstable nucleus decays without capturing any further neutron. The r-process is active in core-collapse supernovae and during the merging of two neutron stars in a binary system, hence it is efficient for very short timescales, while the s-process is active for longer timescales in AGB stars. Here the s-elements produced are mixed to the surface by convection and injected in the interstellar medium by the strong winds typical of these objects. The elements ^{232}Th, ^{235}U, and ^{238}U are produced by the rapid process, but do not decay when the r-process that has formed them is still active, because of their long half-life.

Ideally, if the abundance for example of ^{232}Th can be measured, its ratio (^{232}Th/Xy) to the abundance of a generic stable element Xy tells us how old is the star, provided we know the value of this ratio at birth, denoted as (^{232}Th/Xy)$_0$. In fact, the measured ratio is necessarily lower than (^{232}Th/Xy)$_0$, because ^{232}Th is not produced unless the star is exploding as a supernova, and the decay (with known decay constant λ) slowly transforms ^{232}Th to ^{208}Pb; the difference between (^{232}Th/Xy)$_0$ and (^{232}Th/Xy) can therefore provide the age of the star.

Unfortunately, it is hard to predict theoretically the initial $(^{232}\text{Th}/\text{Xy})_0$ because we need a model for the chemical evolution of the matter out of which the target star has formed. This is subject to several uncertainties and to mitigate this problem, the generic element Xy in the ratio $(^{232}\text{Th}/\text{Xy})$ is usually another r-process element, like europium (Eu). The hypothesis behind this choice is that the ratios of the initial abundances of pairs of r-process elements are somewhat 'universal', meaning that they are the same for all stars. This is partially corroborated by comparisons of abundance ratios between stable r-process elements measured in the Sun, and in a few stars with much lower metallicity. These comparisons show that these ratios in metal-poor (compared to the Sun) stars are in good agreement with the solar values. If this is the case, the initial value of $(^{232}\text{Th}/\text{Xy})_0$ can be deduced from the Sun, after correcting the measured solar abundance of ^{232}Th by the depletion caused by its radioactive decay over 4.57 Gyr (the age of the Sun).

Similarly, the ratio $^{238}\text{U}/^{232}\text{Th}$ is also often employed; given that these two radioactive isotopes have different half-lives (4.5 Gyr for ^{238}U, and 14 Gyr for ^{232}Th) this ratio decreases with time because ^{238}U decays faster than ^{232}Th. It is also argued that theoretical predictions for $(^{238}\text{U}/^{232}\text{Th})_0$ are more trustworthy than for other pairs of r-process elements, because they have very similar numbers of protons plus neutrons.

The use of these radioactive elements to determine the age of stars is generally restricted to objects with a much lower metallicity than the Sun because, due to the presence of nearby stronger lines of other heavy elements, only in metal-poor stars the weak absorption lines of thorium and uranium can be measured properly. Also, at low metallicity we find observationally that r-process elements tend to be proportionally more abundant than other heavy elements compared to the Sun.

7.3 AGE OF STAR CLUSTERS

The determination of the age of star clusters takes advantage of the fact that – as recognized very early on – stars in a cluster are essentially coeval and with the same initial chemical composition. This can be seen very clearly when comparing their observed CMDs[21] with theoretical

---

[21]The MS, SGB, and RGB in the observed CMDs often display an intrinsic width, more obvious in Fig 7.8, due to the unavoidable measurement errors on the observed magnitudes – larger for fainter objects. These errors are equivalent to when we mea-

isochrones – see for example Figs. 7.7 and 7.8 below – and is also confirmed by spectroscopic analyses of stars in individual clusters[22].

The age of a cluster can be determined with the same distance-independent techniques discussed in the previous section, if we have masses and radii of the components of at least one cluster eclipsing binary system, or if we have the mass of at least one post-MS star thanks to asteroseismology. This is the case for a few clusters in the Milky Way, while for some other clusters we can use the Strömgren photometric equivalent of the Kiel diagram. In this latter case, the age of the cluster is found following the same principles discussed below for the case of standard CMDs.

---

sure a temperature with a digital thermometer and obtain slightly different values after the measurement is repeated several times. The result is that the measured brightness of a group of stars all with the same intrinsic luminosity won't be the same for all these stars. These variations, however, are random, towards both higher and lower values, so that their mean brightness is essentially equal to what would be measured when errors are negligible. Measurement errors in different photometric filters are not correlated, meaning that for one filter the measured brightness could be too high, and for the other filter too faint, so that the measured colours are also subject to random variations. In addition, especially for the MS, there is the effect of binary stars too distant for the individual components to be disentangled. They appear brighter than single stars (we measure the sum of the brightness of the two components), spreading vertically the observed sequence.

We can take into account all these effects when comparing theoretical isochrones to the cluster CMDs. For example, along the MS and RGB, we can consider all stars within narrow magnitude bins, and study how their number varies as a function of colour. The colour corresponding to the largest number of objects – called the 'mode' of the distribution – can be then taken as the intrinsic colour of the observed CMD at that magnitude. The same procedure can be followed for the SGB, but due to its horizontal morphology, it is convenient to determine the mode of the magnitude distribution in a given narrow colour bin.

[22]We have discovered during the last two decades that star clusters more massive than about 100,000 solar masses (the sum of the masses of all their stars) are made of stars not exactly with the same initial chemical composition. In a given cluster the initial abundance of a few elements (mainly carbon, oxygen, nitrogen, sodium and helium) changes from star-to-star. However, the pattern of these variations is such that stellar models' lifetimes are mostly unaffected, and cluster CMDs using photometric filters around optical wavelengths conform quite well – apart from very few cases – to isochrones with a single age and initial chemical composition [7, 19]. There is also a handful of globular clusters whose stars display a range of iron abundances and in some case ages, the most notable case being ω Centauri, the largest and brightest globular cluster in the sky.

For the large majority of star clusters – due to their distance – the only accurate observations available are standard CMDs, especially in filters around the optical wavelength range[23].

We have seen both in chapter 4 and also in the previous section about the age of individual stars, that the turn-off region is the most age-sensitive feature of an isochrone. The implication is that, to determine the age of a star cluster from its CMD, the observations must be able to reach stars below the turn-off. This sets an upper limit to the distance of the clusters we can investigate, due to the limited power of our telescopes. At the time of writing, we can just reach the turn-off of the oldest clusters at about the distance of the Andromeda galaxy (about 770 Kpc).

If we know the cluster distance (for example from the parallax of its stars) plus the extinction along the line of sight, and we have measurements of its metallicity (parametrized by [Fe/H]) we need just to pick a set of isochrones of the appropriate initial metallicity, and shift them according to the distance and extinction/reddening to place them on the same diagram as the cluster. A vertical (magnitude) shift accounts for the distance modulus plus extinction, while a horizontal (colour) shift accounts for the reddening. The isochrone that matches the position of the observed turn-off provides the cluster age.

As an example, the top panel of Fig. 7.7 shows a CMD of the Hyades open cluster, the nearest open cluster at a distance of about 150 light-years, for which we have accurate parallax distances to its stars, we know the metallicity (about solar), and the extinction is essentially zero. In this case the observed apparent magnitudes of the cluster stars have been transformed to absolute magnitudes using the measured distances, and the isochrones to be compared with the cluster CMD do not need to be shifted in magnitudes and colours. Notice that the observed CMD contains only a small representative sample of cluster stars, but they are enough to determine its age[24]. The isochrone

---

[23]The chemical composition of the stars is in most cases derived from the brighter cluster stars, generally RGB stars for the globular clusters.

[24]The lack of objects along the SGB and RGB and just two stars in the core He-burning phase are remarkable. These small numbers are certainly due to the small size of the sample of stars in this CMD, but even if we cover the entire cluster, the number of SGB and RGB stars would still be small. This is because of the fast evolutionary timescales of these phases, together with the fact that the whole cluster does not contain many stars. Its total mass (the sum of the masses of all stars) is only of about $400 M_\odot$, about 500 times lower than a typical globular cluster. In addition, for ages below about 1 Gyr, stars along the SGB and RGB do not have an

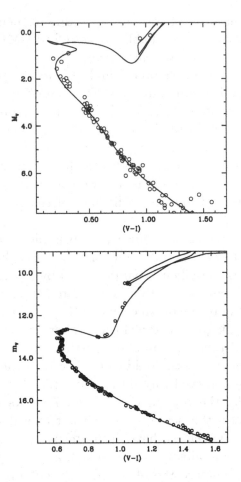

Figure 7.7 Fit of theoretical isochrones to the CMDs of selected stars in the Hyades (top panel, [33] – for this cluster the vertical axis displays absolute magnitudes) and M 67 (bottom panel, [114]) open clusters of the Milky Way. The isochrone matching the Hyades CMD has an age of 900 Myr, whilst the isochrone matching M 67 CMD has an age of 4.5 Gyr. Both isochrones have an initial solar chemical composition [51].

that matches best the turn-off of the cluster provides an age of 900 Myr. Isochrones for younger ages display a main sequence that reaches too

---

electron-degenerate core (see chapter 4): This makes the evolution from the end of the MS until the start of core He-burning comparatively much faster than in stars with electron-degenerate cores.

bright magnitudes compared to observations, while the reverse is true for ages older than 900 Myr.

If we lack independent measurements of distance and extinction[25], or if they are uncertain, the cluster age can be determined using an 'isochrone fitting'.

We select a set of isochrones of the appropriate initial metallicity and start by picking one age. We then shift the isochrone for this age both horizontally (towards larger colours) and vertically (towards fainter magnitudes) in the CMD, until a good match with the observed MS below the turn-off, the RGB (if the RGB is populated) and the core He-burning phase is achieved. If the turn-off and SGB region (the most age-sensitive features) is also matched, this isochrone has the age of the cluster. If not, we repeat the procedure until we find an isochrone that also matches the cluster turn-off. The size (in magnitudes) of the vertical shift corresponds to the apparent distance modulus (the distance modulus plus the extinction A_V in case of the CMDs of Fig. 7.7) whilst the size of the horizontal shift corresponds to the reddening $E(V - I)$ for the CMDs of Fig. 7.7).

Given that reddening and extinction are related – as discussed in the previous chapter – these shifts provide us with estimates of the distance to the cluster, extinction, and reddening. An example is shown in Fig. 7.7 for the solar metallicity open cluster M 67, whose turn-off is best matched with an age of 4.5 Gyr (incidentally, about the same age of the Sun). The distance obtained from the fit is equal to about 2800 light-years, A_V=0.06, and $E(V - I) = 0.03$.

For very old clusters, like the globular clusters in the Milky Way, we can apply the same technique, helped by the properties of old core helium-burning stars. Looking at the CMD of the globular cluster M 68

---

[25]Distances to a cluster with no parallax measurements can be determined by the so-called main sequence fitting technique (MS-fitting), which works like this. If data is available, we can obtain the absolute magnitude as a function of the intrinsic colour of a 'template' MS for the cluster [Fe/H], using real stars with accurate parallaxes and zero or known extinction. To avoid any effect of the likely unknown age of these objects, we should consider this template MS at absolute magnitudes larger (fainter) than about 5.5-6.0 (if we use the V filter), well below the turnoff for ages up to 14 Gyr or more. If the cluster extinction is known and its CMD is corrected for, the vertical shift required to match the template MS to the cluster one, provides the unknown cluster distance modulus. Distances derived with this MS-fitting technique were used to determine in 1997 the globular cluster ages I mentioned in chapter 1 [41], that confirmed the ages found in my works with Achim Weiss and Scilla Degl'Innocenti.

in Fig. 7.8, we cannot fail to notice a roughly horizontal band of stars at apparent magnitudes slightly brighter than 16. A similar feature, albeit less extended in colour, is also seen in the CMD of NGC 6397 shown in Fig. 7.8. This component of the CMDs of all globular clusters is named horizontal branch (HB), and it was first discovered in 1927 by Paul ten Bruggencate. Stellar evolution theory tells us that stars populating the HB are in the core helium-burning phase[26], and the solid lines matched to the lower envelope of the HB of these clusters are theoretical zero age HBs (ZAHBs). They mark the position where models of different total mass begin the proper core He-burning after the ignition of helium in the electron degenerate core at the end of the RGB phase (see chapter 4). The evolution of HB stars is predicted to start from the faint edge of the observed HB, and move slowly towards brighter magnitudes after performing more or less extended loops in the CMD. The blue end (the point with the lowest value of the colour) of the ZAHB corresponds to the smallest possible HB mass, equal to the mass of the helium core at the tip of the RGB; moving towards redder (higher values) colours the mass of the core He-burning models increases steadily. The absolute magnitudes along the ZAHB are independent of age (see Fig. 7.9) for stellar populations older than a few Gyr, and are affected only by the initial chemical composition.

How to explain the different shapes of the HB in globular clusters is an open problem, deserving a long discussion. Suffice here to say that the average colour and the colour extension of the HB change from cluster-to-cluster, in a way that we are not able to predict theoretically yet. If we take the model ZAHB shown in Fig. 7.8 as a guideline, the HB of several clusters appears as a red stub close to the RGB, in others the HB is populated along the horizontal part of the theoretical ZAHB, and yet in other clusters HB stars populate the almost vertical blue side of the ZAHB. There is an observed average trend of the general morphology becoming redder with increasing metallicity, but there are many clusters that do not conform to this trend.

The independence of the ZAHB brightness on age for old stellar populations helps the isochrone fitting when the cluster distance is not known, especially in case the HB is populated along the horizontal

---

[26]Hoyle and Schwarzschild 1955 calculations correctly identified HB stars in globular clusters as objects during the core helium-burning phase. Attempts to investigate the details of their evolution started with the work by Shinya Obi in 1957 [81], and culminated with the first realistic calculations by Iben and John Faulkner, published in 1966 [61].

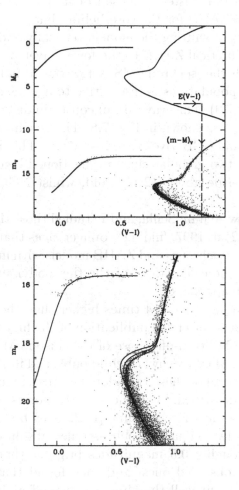

Figure 7.8 *Upper panel:* Fit of a theoretical isochrone up to the tip of the RGB and the corresponding ZAHB to the CMD of the Milky Way globular cluster NGC 6397 [92]. The solid lines display the ZAHB and a 13.5 Gyr isochrone for the cluster metallicity, without (at the top of the diagram) and with (matched to the cluster) the correction for the reddening and distance modulus plus extinction [89]. Dashed lines and arrows show the horizontal and vertical shifts ($E(V-I)$=0.24, and apparent distance modulus $(m-M)_V = 11.96$) applied to the whole isochrone and the ZAHB to match the cluster CMD. *Lower panel:* Fit of theoretical isochrones and ZAHB (solid lines) for 10, 11, and 12 Gyr to the CMD of the globular clusters M 68 [16].

segment. For these old clusters, we use isochrones only up to the tip of the RGB, and the ZAHB for the core helium-burning phase. We then fit simultaneously (by shifting the models vertically and horizontally in the CMD) the theoretical ZAHB to the lower envelope of the observed HB, together with the isochrone MS below the turnoff and the RGB (that in very old populations is insensitive to age, as seen before and shown also in Fig. 7.9). This way we can constrain distance plus extinction and reddening, as shown in Fig. 7.8. The isochrone that matches also the turn-off and SGB gives the cluster age. This is what I did in this figure for the two globular clusters mentioned before: The cluster M 68 turns out to be about 12 Gyr old, whilst NGC 6397 is about 13.5 Gyr old.

This is also how Achim, Scilla, and I studied those three old clusters (M 15, M 68, M 92) in 1997, finding 'younger' ages than previously determined. We used isochrones and ZAHB models with improved physics inputs, especially the calculations of the thermodynamical properties of the gas (see the appendix).

These 'modern' ages are 2–4 times higher than the earliest results obtained in the 1950s, after the publication of the first two CMDs that reached below the MS turn-off, those of the clusters M 92 [3], and M 3 [111]. Sandage and Martin Schwarzschild published in 1952 the first determination of the age of these two globular clusters, using isochrones obtained, as they described, by connecting the points reached by models of various masses at the same time [112]. They first determined the clusters' distances by comparing their assumed absolute magnitude for the HB stars – roughly 0.5 magnitudes brighter than what we find today – with the observed ones, and then found that their 3.5 Gyr isochrone seemed to fit well the clusters' turn-off and MS. Using the same distances, in 1955 Hoyle and Schwarzschild determined an age of 6.2 Gyr from their own stellar models, a result almost identical to what found one year later by Haselgrove and Hoyle (6.5 Gyr for M 3) with their new calculations [46]. All these theoretical models assumed an initial helium abundance much lower than the 25% mass fraction coming from Big-Bang nucleosynthesis, because it wasn't clear yet how much helium was produced during the cosmological nucleosynthesis. Also the treatment of hydrogen burning – among other physics inputs – needed improvements, as shown four years later by Hoyle [54], who also reconsidered the absolute luminosity of stars on the HB used to determine the distance to M 3 and M 92. His conclusion was that Milky Way globular clusters had to be older than 10 Gyr. These early – and for that time ground-breaking – studies, paved the way to the following investigations about the ages of globular clusters and the constraints they set on cosmological models.

7.3.1 Vertical and Horizontal Parameters

In principle, we can also determine the age of old star clusters, like the Milky Way globular clusters, without any isochrone fitting, and without the need to estimate their distance, extinction and reddening, as sketched in Fig. 7.9. These simple techniques rely on the fact that at these old ages the magnitude of the ZAHB and the position of the isochrone RGB in the CMD are both largely insensitive to age, as shown in the figure: When the age increases, only the turn-off position (and the SGB) changes, its colour becoming redder, and its brightness fainter. Quantities like the vertical difference in magnitude between the turn-off and the horizontal part of the ZAHB (ΔV), or the horizontal colour difference between the turn-off and the RGB ($\Delta(V - I)$ if the $(V - I)$ colour is used in the CMD, or $\Delta(B - V)$ if $(B - V)$ is used) can then be employed to determine a cluster's age [113, 115, 124]. An age increase makes the value of ΔV larger (because of the fainter turn-off relative to the brightness of the ZAHB) and the value of $\Delta(V - I)$ (or $\Delta(B - V)$) smaller (because of a redder turn-off compared to the position of the RGB). A comparison of the value of ΔV (or $\Delta(V - I)$) measured from the CMD of an old cluster with the values calculated from theoretical ZAHB and isochrones, provides a straightforward determination of its age. Moreover, the measured values of these vertical and horizontal parameters are independent of the distance, extinction, and reddening, because these quantities cancel out when we measure differences of magnitudes or colours. Even better, the effect of metallicity on the relationship between the vertical (or the horizontal) parameter and age is small because a change of metallicity at fixed age changes the ZAHB luminosity (or the RGB colour) and the turn-off luminosity (or its colour) by almost exactly the same amounts.

A requirement to exploit the full power of ΔV is that the cluster HB must be populated in the horizontal part, and not exclusively in the blue, slanted tail (see Fig. 7.9), otherwise the measured value of ΔV depends on the chosen reference point along the HB. In this latter case, a correction for the reddening is needed to calculate a theoretical

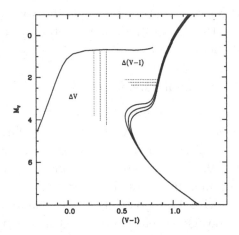

Figure 7.9 Sketch of the 'vertical method' (the difference ΔV) and the 'horizontal method' (the difference $\Delta(V - I)$) for the determination of globular cluster ages. Three isochrones of different ages (8, 10, and 12 Gyr) and the same initial chemical composition, covering the MS, SGB and RGB, are shown together with the corresponding ZAHB. When the age increases, the ΔV increases and $\Delta(V - I)$ decreases.

calibration of ΔV with age that takes the ZAHB magnitude at exactly the same intrinsic colour of the observed HB[27].

7.4 AGES FROM SURFACE CHEMICAL ABUNDANCES

I have mentioned before that it is very hard to determine the age of RGB stars when measurements of mass and radius are not available, because their position in a CMD or a Kiel diagram is a poor age diagnostic. Additional independent constraints on the age of these objects are therefore welcome, to use perhaps in conjunction with CMDs and Kiel diagrams.

---

[27]This also means that the vertical parameter can be only applied when the V filter is used in the CMD because the ZAHB is never horizontal in other filters. More in general, the exact shape of the whole CMD depends on the photometric filters used, for the brightness of a star changes with the wavelength approximately as a blackbody. An object of low T_{eff} will be typically brighter when observed in the infrared (long wavelengths) compared to the ultraviolet (short wavelengths) part of the spectrum, while the opposite is true for a star of high T_{eff}.

As discussed in chapter 5, during the early RGB evolution the first dredge-up changes the surface abundance of C and N (and also He, but its abundance cannot be measured in these stars). More precisely, C decreases and N increases compared to the initial values, and the abundance ratio (C/N) decreases. For our purposes, the 'good' feature of the dredge-up is that at a given metallicity the (C/N) ratio decreases more with decreasing age of the star: The younger the star, the larger the decrease of the (C/N) ratio after the dredge-up [107]. This prediction is confirmed by observations of RGB stars in star clusters with age determined from the CMD.

We can therefore use the difference between the measured (C/N) and its initial value to determine RGB stars' ages. To apply this method – that can be in principle also used to determine the age of star clusters from their RGB population – we first need to know the initial (C/N). The value measured in the Sun is a good approximation, at least for stars in the Milky Way.

We then need to establish the initial metallicity, because the variation of (C/N) due to the dredge-up also depends on Z. This is possible by measuring, for example, the abundance of Fe that can be translated to Z by knowing whether the star belongs to the disk or the halo of the Milky Way, as briefly discussed in chapter 6[28].

Finally, we need to ensure that the target stars have completed the dredge-up, but are fainter than the RGB bump luminosity (see chapter 5), because at the brightness of the RGB bump another mixing process – much more uncertain to model theoretically – starts again to decrease the (C/N) ratio. To this purpose, we can take advantage of the values of g and T_{eff}, always measured together with chemical abundances. A comparison of the measured surface gravity and effective temperature with stellar models for the metallicity of the star[29] allows us to get a good idea of whether the measured (C/N) truly reflects the effect of the dredge-up, independent of the unknown age of the target.

This method has been applied to large samples of stars from spectroscopic surveys of the Milky Way [75, 74]. The precision of individual

---

[28]As already mentioned, the effect of gravitational settling, radiative levitation and mixing due to rotation are almost completely wiped out when the star starts its RGB evolution, with very deep convective envelopes. The measured abundance of Fe truly reflects the initial one.

[29]It is essentially a comparison with evolutionary tracks in the Kiel diagram. We know at which point on tracks of different masses the dredge-up has been completed, and the location of the RGB bump.

ages is not high, because in the age range between 1 and 14 Gyr the predicted post dredge-up values of (C/N) vary just by a factor of 4 at the lowest metallicity, and by less than a factor of 3 at solar metallicity. This range has to be compared to typical measurement errors on (C/N) on the order of a factor 1.3–1.4.

Another technique based on measurements of surface chemical abundances and used to determine the age of nearby young star clusters, is the so-called lithium depletion boundary (LDM) method [119]. This method requires the knowledge of the distance to the target, and is based on the following results obtained from calculations of stellar models:

- Stars with mass below about $0.3M_\odot$ are always completely convective, both during the PMS and the MS;

- The central temperatures of PMS stars increase with time during their evolution along the Hayashi track (see chapter 4). Eventually, the temperature in the core reaches 2.5 million K, high enough for lithium-burning to be efficient. Lithium (Li) nuclei capture protons and produce helium, but the initial abundance of Li in stars is typically so low (mass fractions on the order of $10^{-8} - 10^{-9}$) that there is no substantial amount of energy produced. In these fully convective objects Li-burning depletes Li uniformly from the centre to the photosphere, because convection mixes the whole star very fast.

- More massive stars reach the Li-burning temperature earlier, at brighter luminosities.

Let's suppose to have a self-repairing robotic space telescope that observes continuously a newly formed star cluster, focusing on the region of the CMD populated by fully convective stars. After about 10 Myr the telescope will record the surface abundance of Li drop to zero in the $0.3M_\odot$ stars, because at this age these fully convective objects reach the Li-burning temperature in their cores. The small amount of initial Li is burned very rapidly, while the less massive and fainter stars still retain their initial Li content. With increasing age, stars of progressively lower masses burn Li and the magnitude threshold between stars with surface lithium and stars without – the LDM – becomes increasingly fainter. Beyond an age of about 250 Myr, the magnitude of the LDM no longer changes, because we have reached the lower mass limit for a star (slightly below $0.1 M_\odot$); objects with mass below this limit are brown dwarfs, which never reach the Li-burning temperature.

In practice, if the cluster distance and metallicity are known, the absolute magnitude of the brightest fully convective star that still shows Li at the surface – the brightness of the LDM – can be translated to a cluster age, using the relationship between LDM magnitude and age as predicted by stellar models for the appropriate metallicity (this method obviously works only for ages between about 10 Myr and about 250 Myr).

That's it for this chapter, which has delved into several methods to determine the age of single stars (starting with the Sun) and star clusters, young and old. But what about 'dead' stars like white dwarfs? Can we tell from observations how long ago their cores stopped producing energy and new chemical elements? Can we use these stellar corpses to find alternative ways to determine the ages of stellar populations? Do we learn something new by doing this?

GLOSSARY

Half-life: Given a sample of unstable nuclei, their half-life is the time taken for half the nuclei to transform ('decay') into a different element.

Isochron: Line of constant age in a diagram showing on the horizontal and vertical axes abundance ratios between two different pairs of elements (at least one of them being unstable).

Radioactive decay: A nucleus with the *wrong* number of neutrons is unstable, and decays to a different element, transforming a neutron to a proton with the emission of an electron and an antineutrino, or via other processes.

FURTHER READING

S. Cassisi and M. Salaris. *Old Stellar Populations*. Wiley-VCH, 2013

J. J. Cowan, F-K. Thielemann and J. W Truran.Radioactive Dating of the Elements. *Annual Review of Astronomy and Astrophysics*, 29: 447, 1991

G. B. Dalrymple. *The Age of the Earth*. Stanford University Press, 1994

D. R. Soderblom. The Ages of Stars. *Annual Review of Astronomy and Astrophysics*, 48:581, 2010

CHAPTER 8

Crazy Diamonds

In the mid-1990s I left Italy for good and went to work with the team lead by Jordi Isern at the Centre for Advanced Studies in Blanes, Catalonia, one of the most beautiful and prosperous regions in Spain. Blanes is a lovely town, about an hour by train north of Barcelona, nestled on the Mediterranean coast (on the 'Costa Brava') with coves and beaches surrounded by mountains, and a mild climate even in winter (I still remember that the apartment where I lived had no heating). Even in the touristic season, the town had a relaxed atmosphere, family-friendly, without the bustling nightlife of the nearby Lloret de Mar.

The institute stands on a hill at the edge of the town and at the time housed, together with the small group of astrophysicists, a research group in marine biology. When I arrived in the winter, Blanes was quiet, sleepy, with closed or semi-deserted bars and restaurants. I used to have dinner in a pizzeria called *Paparazzi*, named after the owner and cook, Jorge Paparazzi, an Argentine of Italian origin with an explosive character. I had dinner at *Paparazzi* almost every day until spring, often the only client in the winter evenings. Once a week I went to the local cinema, with a double-feature on offer every day (titles changed once a week). This was very useful to master the Spanish language. To my great regret, I never managed to learn properly Catalan, even though I was able to read it with ease. I also tried to keep up with my running (although I stopped competing in road races, half-marathons and marathons after I left Italy) with evening training along the promenade or the hills surrounding the village.

One year after I arrived, the astrophysics group moved to the newly created Institute for Space Studies of Catalunya, in Barcelona, a beautiful and vibrant city, but I mourned the move away from the enchanting Blanes.

In Catalonia I learned about white dwarf stars and white dwarf physics, well beyond the basics taught at the stellar evolution course I took as an undergraduate student. I spent lots of time in the library reading classic papers about these objects and the very interesting physics going on in their interiors, although I was at first a bit hunted by the following remark, reportedly made by Icko Iben in the mid-1980s: *Any fool can make a white dwarf.*

Actually the structure and evolution of a white dwarf (WD from now on) is simpler than stars in other phases, but accurate modelling is by no means an easy task, because of the extremely high densities and comparatively low temperatures of WD interiors. As we already saw in chapter 3, it was clear since the mid-1930s that in a WD electron degeneracy sets in and provides the pressure to maintain hydrostatic equilibrium. As discussed by Leon Mestel in 1952, WDs cannot be electron-degenerate and at the same time be powered by nuclear reactions, otherwise they would have very high central temperatures and reach the 'normal' equilibrium state at much larger radii, with the electron degeneracy removed [77]. There were now two questions to be addressed. How do WDs form, and how do they evolve in time?

8.1 WHITE DWARF EVOLUTION

I mentioned in chapter 3 Russell's view that the Sun would become a WD at the end of core hydrogen-burning, implying that WDs are the result of hydrogen exhaustion of a MS star of the same mass. Following improvements in the understanding of stellar physics and evolution, and the first determinations of open and globular cluster ages, Öpik put forward in 1953 the idea that WDs might be born out of condensation of what he called 'meteoritic material'. This material was supposed to be made of just heavy elements and no hydrogen, produced by some mechanism that separated hydrogen from metals in the interstellar medium, during the early stages of the Milky Way evolution. This idea came after the realization that stars with masses as low as the known WDs (below one solar mass) had very long MS lifetimes, and couldn't have possibly completed yet hydrogen burning in their cores, assuming a 5–6 Gyr age of the universe from the ages derived at the time for the oldest clusters. The following year, Öpik speculated that possibly some WDs might come from the evolution of stars with normal initial composition, which after the MS lose the outer layers around the helium core, in an episode of violent ejection. The remaining helium core then undergoes nuclear reactions that produce mainly magnesium (the details of stellar nucleosynthesis after the MS were still being worked out at the time) before electron degeneracy sets in.

In 1956 Armin Deutsch discussed the possibility that what we now call AGB stars might be the progenitors of WDs after losing their envelope. He based his idea on measurements of their mass-loss rates, and number counts of WDs and AGB stars [31]. The study of star clusters with improved stellar evolution models confirmed this idea that WDs must have originated from stars that lost a substantial amount of their gas after the MS. An example is the 1965 investigation of the open clusters Hyades and Pleiades by Larry Auer and Neville Woolf [4]. The mass of these clusters' WDs (inferred from their position in the CMD and the relationship between mass and radius) is typically below $1M_\odot$, while the mass of the stars at the MS turn-off is of about 2.2 $M_\odot$ for the Hyades, and about 5 $M_\odot$ for the Pleiades (they got these values from the isochrones that fit the observed CMDs). This means that along the MS there are stars with a mass comparable to the cluster's WDs (smaller than the mass of turn-off stars) and, as a consequence, the WD progenitors must have been initially more massive than the objects at the turn-off.

On the theoretical side, the attention focused on models of stars that have exhausted helium in the core and try to raise their core temperature to start carbon burning; it was realized that below a certain value of the initial mass (see the discussion in chapter 4) the helium-exhausted core becomes electron-degenerate, and the onset of strong neutrino emission that takes away energy from the star prevents the ignition of carbon [125, 83]. These are the AGB models described in chapter 4, with an electron-degenerate core surrounded by a helium-burning shell, and a more external hydrogen-burning shell. A strong mass loss would remove on short timescales almost all the envelope around the core, and the models cross the HRD (or the CMD) at constant luminosity – the so-called post-AGB evolution – towards increasing T_{eff} and decreasing radii (see Fig. 4.3). When the effective temperature of the models reach about 30,000–40,000 K, the energetic photons released from the photosphere are expected to heat up the surrounding gas lost during the AGB, which becomes visibile at optical wavelengths, and produces a 'planetary nebula', the beautiful display shown in Fig. 8.1. The term planetary nebula is actually a misnomer

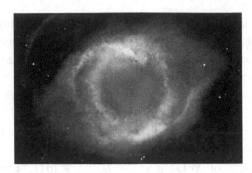

Figure 8.1 Helix nebula (a planetary nebula) in the constellation of Aquarius, at a distance of about 650 light years (Allexxandar/shutterstock).

because these objects have obviously nothing to do with planets, but it has been maintained for historical reasons. They were first discovered during the late eighteenth century, and with the telescopes of the time they appeared to have a disk-like structure, looking similar to Uranus and Neptune, hence the association with planets. Observations of cores (the central stars) of planetary nebulae with radii close to those predicted by the WD mass-radius relationship reinforced the connection between AGB, post-AGB, and WD stars.

The HRD crossing stops after 10,000–20,000 years when the models settle on the equilibrium radius appropriate for their mass, and the proper WD evolution starts. About 99% of the mass of a WD model is in an electron-degenerate core made of C and O (plus the small fraction of the other metals present at birth)[1] with a uniform temperature, because electron conduction (see chapter 4) redistributes energy very efficiently across the degenerate core. The upper 1% of the WD mass is made of the outer part of the helium core formed at the end of the MS, which has grown throughout the post-MS evolution thanks to hydrogen-burning in the surrounding shell. The outermost WD layers are an almost evanescent envelope (mass at most equal to around 0.0001 times the mass of the WD) made mainly of hydrogen, with

---

[1]Stars with an initial mass between $6-7M_\odot$ and about $10M_\odot$ are expected to make WDs made of a O-Ne core, with masses probably around 1.1–1.3 $M_\odot$. After degeneracy sets in at the end of core carbon-burning, these stars fail to ignite the next chain of fusion reactions and experience a strong mass-loss from the surface. The expected number of O-Ne WDs is small compared to the number of C-O WDs, because of the narrow mass range of the progenitors.

essentially the initial chemical composition of the progenitor, modified only by the dredge-ups (see chapter 5). These non-degenerate helium and hydrogen envelopes are unable to support hydrogen and helium nuclear burnings[2].

Spectroscopy shows that a fraction of WDs does not show evidence of a hydrogen envelope, due to differences in the exact moment when their progenitors terminate the AGB evolution due to the surface mass loss[3].

The relationship between the final mass of a WD and its initial progenitor mass is difficult to predict theoretically, due to the uncertainties in modelling the efficiency of mass-loss during the RGB and AGB phases. We will briefly see in the next section what we do to determine this relationship, whose general behaviour is shown in Fig. 8.2. All initial masses up to about $2M_\odot$ (including the Sun) produce WDs with roughly the same mass, equal to about 0.55-0.56 $M_\odot$. The WD mass then increases steeply when the initial mass goes from 2 to $4M_\odot$, followed by a more gentle increase for higher initial masses. The maximum mass for C-O WDs seems to be around 1 $M_\odot$, meaning that mass loss during the AGB phase prevents WDs from reaching the Chandrasekhar mass.

At the beginning of the WD evolution, gravitational settling (see chapter 4) makes the small fraction of metals in the hydrogen and helium layers sink very fast down to the edge of the electron-degenerate core (and the helium from the outermost hydrogen-dominated layers, if present, settles in the underlying helium envelope) as first shown by Evry Schatzman in 1945[4]. The non-degenerate outer layers are therefore made basically of a layer of pure helium, surrounded by a much thinner pure hydrogen layer. The core temperature at this stage is still

---

[2]There can be some residual hydrogen burning during the early WD evolution, provided the mass of the remaining hydrogen layers is higher than about 0.0001 times the total WD mass, although the precise value does depend on the mass of the WD and the metallicity of the progenitor.

[3] White dwarfs with hydrogen envelopes are assigned spectral type DA, and DB is the spectral type of WDs with just the helium envelope. There are relatively few WDs with more complex spectra, showing a mixture of H and He, and sometimes also small amounts of metals. These objects are denoted with spectral types DO, DQ, DZ, DC, DAB, DAO, DAZ, DBZ. Accretion of metals from interstellar matter and convective mixing between H and He envelopes play a major role in producing this variety of photospheric compositions.

[4]Surface gravities of WDs are typically 1000- 10,000 times higher than the solar gravity.

very high, equal to a few hundred million Kelvin[5], but cannot support any nuclear reaction. As summarized in chapter 4, WD models predict an evolution that is an indefinite cooling at almost constant radius. The C-O core is the energy reservoir, the outer non-degenerate layers control the rate of energy outflow, and the luminosity and core temperature decrease. The speed at which the energy is radiated away depends on the opacity[6] of the outer layers: The higher the opacity, the slower the release. It is a bit like having a hot ball wrapped in a blanket; if I pick a thin blanket, the ball will cool down faster than in case I use a very thick cover.

Leon Mestel pioneered the theoretical investigations about WD evolution in his classic 1952 paper titled *On the theory of white dwarf stars*. He provided the first relationship – the 'Mestel law' – between luminosity, mass, chemical composition, and age of a WD, without calculating full detailed models, in the assumption that WD cores are made of degenerate electrons and a perfect gas of fully ionized atoms (I will call them ions). Armed with his relationship, Mestel derived the ratio of the ages (from the onset of their formation) with respect to the unknown value of the mean atomic weight of the elements that make up the cores of eight WDs. In 1954 Öpik obtained ages of at most 4 Gyr for the eight WDs studied by Mestel, considering a core composition of magnesium.

The seminal Mestel study was followed between 1960 and 1970 by a number of fundamental works that refined our description of the physics of WD cooling [7]. During the same time, stellar evolution calculations improved our understanding of the origin of WDs, and the chemical composition of their cores.

Our current picture of the evolution of WDs can be sketched as follows. The source of the energy lost from the surface is provided essentially by the kinetic energy of the ions, because due to the quantum mechanical degeneracy, electrons cannot lose energy. The loss of the ions' energy causes a steady decrease in the temperature of the core, and of the luminosity L as well; given that the radius R stays

---

[5]The temperature for the ignition of carbon-burning is around 600 million Kelvin.

[6]We have seen in chapter 4 that the opacity of the stellar matter is simply a measure of the resistance posed by the gas to the outgoing flux of photons.

[7]I refer here especially to works by Alexei Abrisokov, David Kirzhnits, Attay Kovetz and Giora Shaviv, Mestel and Melvin Ruderman, Edwin Salpeter, Hugh van Horn.

approximately constant[8] during the evolution, also T_{eff} keeps decreasing because $L = 4\pi R^2 T_{eff}^4$.

The first stages of the cooling (as the WD evolution is often called), down to luminosities of about one-tenth of the solar luminosity, are fast, because due to the still high temperatures of the C-O cores and their high densities, a large number of neutrinos is produced (see chapter 4). As a consequence, a substantial amount of energy is taken away from the core without contributing to the energy reservoir.

With decreasing temperature the random motion of the ions (that have a positive electric charge) become progressively slower, the electrostatic interactions among particles become stronger[9], and the ions start to behave like a liquid, with much less freedom of movement within the core. When the core temperature (and luminosity) decreases further, the ions in the core experience a transition from liquid to solid ('crystallization'), whereby they arrange themselves in a periodic lattice structure to minimize the strength of the electrostatic interaction, and their motion is reduced to oscillations around their positions in the lattice. Ions behave now like atoms in a crystal, which are organized in a regular pattern repeated throughout the whole sample.

Crystallized C-O WDs are often compared to diamonds, because diamond is a form of solid carbon with atoms arranged in a cubic structure. However, a crystallized WD core is made of ions held in position by their mutual repulsion, while in a diamond, the carbon atoms are held in position by bonds formed by their electrons. Models show that the density at the centre of a WD with a mass equal to $0.55 M_\odot$, roughly the expected final WD mass of the Sun, is about one million times higher than the density of diamond.

When the ions undergo the transition from liquid to solid, extra-energy is released, proportional to the temperature of the core and the number of crystallizing ions, and the cooling process slows down. This energy is the 'latent heat', and is analogous to what happens when liquid water transforms into ice. When heat is subtracted from liquid water, the water molecules slow down; as T decreases down to 273 K (equal to zero degrees Celsius) water turns into ice at constant temperature, and releases the extra-energy the molecules had when they could move more freely.

---

[8]Detailed calculations show that the radius of WD models actually shrinks very slowly and by small amounts during the evolution.

[9]The ions can no longer be considered a perfect gas.

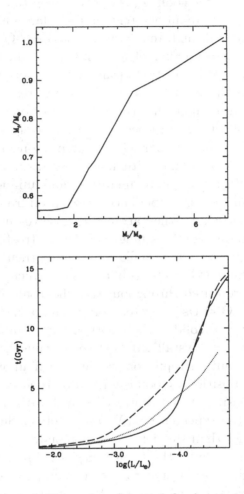

Figure 8.2 *Top panel:* Example of initial-final mass relation for C-O white dwarfs. The vertical axis displays the final WD mass, and the horizontal axis the corresponding initial mass of the progenitor (both in solar mass units). *Bottom panel:* Cooling time (vertical axis) as a function of luminosity (horizontal axis) for 0.55 (with H – solid – or He atmospheres – dotted line) and $1.0 M_\odot$ (H atmosphere – dashed line) WD models [102].

Another important process is in action during crystallization, investigated in detail during the late 1980s and 1990s by the group I worked with in Blanes and Barcelona [62, 106]. In a nutshell, the chemical composition of the C-O core in the gas and liquid phase cannot be the same as in the solid phase, because during the crystallization process – that starts from the denser centre and propagates towards the less dense edge of the core with decreasing temperature – oxygen ions tend to be displaced towards the centre, and carbon ions are pushed towards the edge of the core. As a result, the chemical composition of the core changes; the displacement of the heaviest element in the core (oxygen) towards the centre releases gravitational potential energy (it is qualitatively like having a small contraction of the core, which makes it more condensed) that adds to the latent heat, further slowing down the cooling. When crystallization is completed, the cooling speeds up, becoming increasingly faster because of the limited energy of the ions, which can only oscillate around their positions in the crystallized core.

Figure 8.2 shows the evolution of the luminosity as a function of time since the beginning of cooling, of WD models with different masses – 0.55 and 1.0 $M_\odot$ – and the same composition of the envelope (pure helium layers with mass equal to 1% of the WD mass, and hydrogen layers with mass equal to 0.01% of the WD mass). The steeper part of the curve (slower decrease of the luminosity with time) is a signature of the crystallization advancing through the core[10]. The same figure shows the effect of removing the hydrogen layers on the lower mass model. The evolution is accelerated because the opacity of hydrogen is higher than the opacity of helium (hydrogen makes a thicker blanket than helium) and the heat is released at a faster rate in this model. As a consequence, crystallization starts earlier than in the calculation with an hydrogen envelope, and the cooling speed is generally faster.

The well-defined relationship between cooling time and luminosity (the fainter the luminosity, the higher the age), provides a clock to determine the ages of these stellar remnants.

Figure 8.3 shows the HRD of evolutionary tracks of C-O WD models ('cooling tracks') for various masses: They do not look particularly exciting, being roughly straight lines corresponding to an approximately constant radius. At a given effective temperature more massive models

---

[10] Crystallization starts at higher luminosities – meaning also higher temperatures – in the more massive and denser (because of the lower radius and higher mass) model.

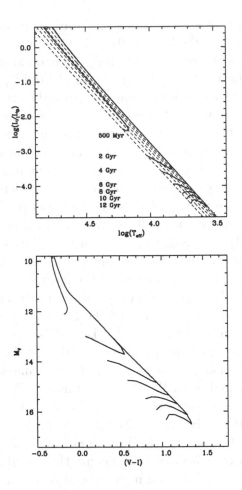

Figure 8.3 *Top Panel:* HRD of WD cooling tracks (dashed lines) with masses between 0.53 and 1.0 $M_\odot$ and isochrones (solid lines) for the labelled ages. *Bottom Panel:* The same isochrones but in an optical CMD.

are fainter, because of their lower radius. The same figure shows the extension of isochrones to the WD phase (both in the HRD and in a CMD using the V and I photometric filters) whose calculation follows the same recipes discussed in chapter 4. When the isochrone age t_i is fixed, we need to pick one point on each cooling track where the sum of the model cooling age plus the progenitor age until WD formation is equal to t_i. Once these points are connected, we get our isochrones extended to the WD regime. The ingredients to calculate a WD isochrone are therefore WD cooling models, plus an initial-final mass relation to link the WD mass to its progenitor initial mass, and the lifetime of the progenitor until the WD stage.

The shape of WD isochrones is interesting. For the most part, they follow lines of constant radius, corresponding roughly to a constant WD mass, and only at the faint end they display a turn towards higher T_{eff}, denoting the presence of increasingly larger WD masses. This behaviour can be understood when we notice that at each luminosity of the WD isochrone $t_i = t_{cool} + t_{prog}$ where t_{cool} is the cooling time of a WD model and t_{prog} the lifetime of its progenitor. We have seen that t_{cool} is very short at the bright end of the cooling sequence (see Fig. 8.2), and practically negligible compared to the age of the isochrone, unless we have chosen a very young age on the order of tens of millions of years or less. As a consequence, t_{prog} will be approximately equal to t_{iso} and roughly constant at these luminosities; but constant t_{prog} means constant progenitor mass, hence a constant WD mass

With decreasing luminosities t_{cool} becomes a more sizeable fraction of t_i, and more massive WDs coming from higher mass progenitors with shorter t_{prog} do appear. However, as long as the mass of the progenitor is below about $2M_{\odot}$, the WD mass evolving along the isochrone hardly changes, because of the approximately constant value of the final WD mass for progenitors below about $2M_{\odot}$ (see Fig. 8.2). A clear increase of the WD masses along an isochrone appears at its faint end, which turns towards larger T_{eff} values (smaller radii). The luminosity of the faint end of WD isochrones decreases with increasing age, because of the longer cooling times that make the WD models increasingly fainter.

In general, the initial chemical composition of the progenitors affect the relative abundances of C and O in the core of the WD models, which in turn have an effect on the cooling times. Also the initial-final WD mass relation, and the mass thickness of the helium and hydrogen envelopes can in principle be affected by the initial composition of the progenitor, but the theory is still unable to make firm predictions on this.

8.2 AGES OF WHITE DWARFS

The best way to determine the age of individual WDs – meaning their cooling age, the time since they have entered their cooling sequence – is through the Kiel diagram. Spectroscopy provides us with the composition of the photosphere – either H or He – plus T_{eff} and surface gravity g, which can be compared to predictions from WD evolutionary tracks with the appropriate envelope composition. An example is shown in Fig. 8.4, where the data for WD 1633+572, a WD with a helium envelope lacking the surrounding hydrogen layers [40], are plotted together with He-envelope WD tracks [102]. The tracks are practically horizontal in this diagram because the evolution is at a constant radius (g is proportional to the mass divided by the square of the radius) and age increases moving towards lower T_{eff} values. WD tracks of different masses run parallel to each other, and this allows us to determine both the mass and the age of individual WDs. The measured g determines the mass of the observed WD, and T_{eff} fixes the age. In the case of WD 1633+572, its mass is between about 0.54 and 0.61 $M_{\odot}$, and the age between 3 and 3.8 Gyr.

If we can determine the cooling ages of individual WDs as just described, we can also constrain from observations the initial-final WD mass relationship, which is hard to predict from calculations of stellar models. The best way to do this is as follows.

Through spectroscopy of the brightest WDs in nearby open clusters and globular clusters we can measure cooling ages and masses like in Fig. 8.4. The differences between the cluster age measured from the MS turn-off (see chapter 7) and the cooling ages of its WDs give the ages of the progenitors until the WD formation. These progenitor ages can be translated to progenitor masses using results of stellar evolution calculations [108]. A relationship between progenitor initial mass and age until, for example, the end of core helium-burning will suffice. This age corresponds to well over 97% of the lifetime until WD formation, irrespective of the uncertainty on the following AGB evolution.

8.2.1 Star Clusters

The age of star clusters can also be determined from the CMD of their cooling sequences, by using theoretical isochrones that include the WD evolution. On the observational side, we need CMDs that reach the faint end of the WD sequence, because its brightness is the age indicator, as shown in the previous section.

Figure 8.4 displays the CMD of the Milky Way globular cluster NGC 6397, one of the handful of globular clusters with detections of the faint end of the WD sequence to date. The same figure shows an isochrone for the MS-SGB-RGB sequences, and a WD isochrone for

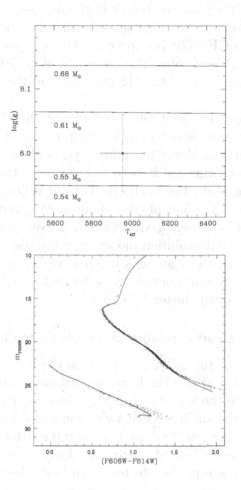

Figure 8.4 *Top panel:* Age determination of WDs from the Kiel diagram. The filled circle displays data (with error bars) for the object WD 1633+572, solid lines show a set of WD evolutionary tracks with the labelled masses (cooling age increases towards lower T_{eff}). More massive WDs evolve at higher g because of their larger mass and smaller radius. *Bottom panel:* Fit of a 13.5 Gyr WD isochrone [102], and same age isochrone from the MS to the RGB [89], to the observed CMD (corresponding to $V - (V - I)$) of the globular cluster NGC 6397 [93].

models with hydrogen envelopes[11]. The observations are optimized to reach faint objects, and this explains the depopulated RGB and the lack of HB stars: They are too bright in the images.

Distance modulus and reddening/extinction have been derived from the fit of the MS-SGB-RGB isochrone to the data (see chapter 7) and applied to the WD isochrone. The same age of 13.5 Gyr is obtained from both the magnitude of the MS turn off, and the faint end of the WD sequence.

It is worth recalling that the faint end of the WD isochrone is populated by the more massive WDs, the progeny of stars that have left the MS as early as about 100 million years after the formation of the cluster. Their cooling age is almost the same as the MS lifetime of stars at the MS turn-off. The isochrone TO and WD ages are therefore probing very different stellar physics: That of MS stars, and the more extreme regime of WD cooling. Checks of their consistency are another powerful test of stellar evolution models, for any discrepancy between turn-off and WD cluster ages tells us that we are still missing some important physics in our calculations. This is beautifully exemplified by the case of the open cluster NGC 6791.

8.2.2 The White Dwarf Population of the Old Open Cluster NGC 6791

NGC 6791 is one of the oldest known open clusters, at a distance of about 13,000 light-years, in the Lyra constellation. Apart from its old age, comparable to the age of globular clusters, another peculiarity of this cluster is its metallicity, about twice the metallicity of the Sun.

Between 2005 and 2008, as part of an international collaboration with colleagues in the United States and Italy, I worked on the determination of the age of this cluster, from both the MS turn-off and the WD cooling sequence, using observations made with the Hubble Space Telescope [11, 10, 9]. The first set of data in 2005 wasn't clear regarding the faint end of the WD sequence, but we already noticed something puzzling. Three years later, with data that reached fainter magnitudes, we got the confirmation that we had a problem. I remember those times vividly, but for the wrong reasons: I was suffering from a slipped disk in the lower spine, that touched the spinal cord and paralysed my left leg for a few weeks, in addition to unbearable pain. My back was presenting me with the ultimate complaint after years of

---

[11]Spectroscopy of some bright cluster's WDs suggest that WDs in globular clusters are all with hydrogen envelopes.

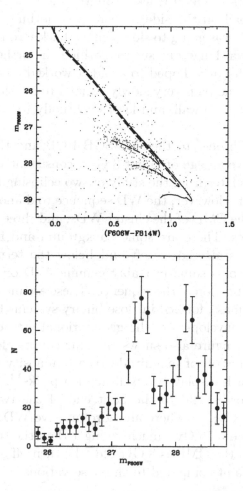

Figure 8.5 *Top Panel:* Hubble Space Telescope CMD (corresponding approximately to V on the vertical axis and $(V-I)$ on the horizontal one) of the WDs in the old open cluster NGC 6791, compared to three WD isochrones (models with hydrogen envelopes) with ages equal to 4, 5.75 (dashed lines) and 8 Gyr (solid line) respectively. The age derived from the MS turn-off is equal to 8 Gyr. *Bottom panel:* Luminosity function (number of stars per magnitude interval as a function of magnitude) of the WDs in NGC6791. The thin vertical line marks the magnitude beyond which the star counts are unreliable.

abuse because of running thousands upon thousands[12] kilometres on tarmac, pavements, athletic tracks, countryside (not to mention basketball and football on the side). There was nothing I could do but hope that the disk was going to slowly move back into place, otherwise I had to undergo a dangerous surgery, which the orthopaedic consultant at my local hospital hoped to avoid. I worked on these new data while fortunately the pain very slowly started to subside, and I began regaining the ability to walk and later eventually to start doing sport again.

The fit of isochrones to the MS-SGB-RGB and the core helium-burning phase gave an age of about 8 Gyr, consistent with the age inferred a few years later from the study of two eclipsing binaries discovered in the cluster. However, the WD sequence told another story. Figure 8.5 displays the CMD of the cluster WDs, and three WD isochrones with different ages. There are some background and foreground contaminant objects scattered to left and below the termination of the cluster sequence, plus some probably genuine WD on the right-hand side of the bright portion of the sequence. These are most likely helium-core WDs of low mass, formed in close binary systems from RGB stars stripped of their envelopes by the gravitational attraction of a companion. The same figure also shows the star counts along the cooling sequence as a function of magnitude (the luminosity function of the cluster WDs) which display two well-defined peaks[13]. The magnitude of these peaks correspond to the faint end of the two younger WD isochrones in the CMD – where all more massive WDs pile-up – with ages equal to 4 and 5.75 Gyr, much younger than the turn-off age. The older isochrone in the CMD is 8 Gyr old (the turn-off age) and reaches magnitudes too faint compared to the observations.

---

[12]Running for example 30-40 kilometres a week including eventually track workout –a reasonable amount (on the short side) for an amateur athlete competing in races up to 12-15 kilometres– for let's say 40 weeks a year, makes already 1200–1600 Km in just one year. Training for marathons requires much longer distances per week, at least in the last months leading to the competition.

[13]You may notice that the numbers of stars in the faintest magnitude bins of the luminosity function are larger than the number of points in the CMD. The reason is that towards the faint end of the observations, there is an increasingly high chance to miss stars in our images. However, we can fairly easily estimate how many stars go undetected as a function of magnitude. The luminosity function shows the numbers of stars in the CMD corrected for the fraction of objects missed by the observations. The thin vertical line marks the limit where the correction is more than 50% of the observed numbers. Corrections larger than this fraction become unreliable.

We had to face a major problem: There are two terminations of the WD sequence, like two populations of different ages. But the cluster does not have two distinct MS turn-offs. Also, the MS turn-off age was too high for all the observed WDs. Realistic variations of the initial-final WD mass relations wouldn't solve the problem, as using models with just helium envelopes, because they cool down faster. A 8 Gyr WD isochrone from models with helium envelopes would reach fainter magnitudes than isochrones for hydrogen envelopes, exacerbating the discrepancy with observations.

While we were writing-up our result, saying that we did not have a solution to this puzzle, our attention was drawn to two papers published a few years earlier by Lars Bildsten with David Hall and Christopher Deloye, respectively. They expanded upon an early idea by Jordi Isern (my boss when I was working in Spain) and Eduardo Bravo. The papers discussed the following: The C-O core of WDs contains a small percentage of neon (Ne – with a nucleus heavier than carbon and oxygen), produced during the previous core helium-burning evolution of the WD progenitors. The mass fraction of this Ne is very close to the initial metallicity Z of the progenitor, equal to about 2% for progenitors of solar initial composition, and much less for globular clusters' WDs. When the core is cold enough to be in the liquid phase, calculations show that this Ne slowly tends to drift towards the centre, and stops only when crystallization sets in. This 'neon gravitation settling' tends (like what happens at crystallization) to displace the heaviest element in the core towards the centre, releasing gravitational energy that slows down the cooling. This makes the termination of WD isochrones brighter at a given age. The higher the initial Z of the progenitor, the higher the amount of Ne in the core and the larger the impact on cooling times. In fact Bildsten and Deloye pointed out in 2002 that the effect of neon settling on WD cooling should be detectable in NGC 6791, whose population has an initial Z about twice the solar value.

No detailed WD models including this additional effect were available at the time, and we concluded the paper stating that Ne settling might be able to explain the discrepancy between the turn-off age and the age of the fainter peak of the WD star counts, those objects that we matched with a WD isochrone of about 6 Gyr. Confirmation of this solution to one of NGC 6791 puzzles came a couple of years later, with a study led by Enrique García-Berro[14] (with whom I had worked already when I was in Jordi Isern group) to which I was invited to collaborate [39]. This work presented complete calculations of WD evolution including also Ne settling, showing that a 8 Gyr old WD isochrone can match the faint end of the cluster WD sequence.

The comparison of turn-off and WD ages in this cluster bore fruits. We have been able to confirm the existence of an additional physics process, that affects WD populations with high metallicity progenitors.

As for the brighter 'termination' of the cooling sequence, the problem is still open. With my colleagues, I proposed that we are seeing binaries made of two WDs, blended together (hence more luminous than single objects) because of their distance. My simulations showed that, assuming an age of 8 Gyr, these binaries would end up exactly where we see this brighter group of WDs in both CMD and luminosity function. This solution requires a high fraction of binary stars in the cluster. An alternative solution proposed by Brad Hansen envisages these objects as the progeny of RGB stars in binary systems, that just missed helium ignition because they were stripped of their envelopes by the gravity of the companions, very close to the tip of the RGB. They would therefore produce massive – with a mass of about 0.5 $M_\odot$ – He WDs (made of the electron degenerate helium core of the RGB progenitor), with a slower cooling compared to a C-O WD of the same mass[15].

GLOSSARY

Crystallization: Process by which the ions in the core of a white dwarf get organized into regular patterns repeated throughout the star.

Latent heat: Energy released when the ions in the core of a white dwarf crystallize.

---

[14]Enrique died in a tragic accident in September 2017, while climbing alone the 'Picos de los Infiernos' ('Peaks of Hell') in the Spanish Pyrenees, three mountains over 3,000 metres high. Only a few months earlier, in June, I finally met him again after several years, at a conference in a lovely village by the sea, in Catalonia, near Blanes.

[15]For a given total mass, a He-core WD will contain more ions than a C-O one, because a helium ion is lighter than a carbon or an oxygen one. Given that the energy available to a WD is the energy of its ions, a larger number of ions mean more energy, and longer cooling times.

FURTHER READING

B. M. S. Hansen and J. Liebert. Cool White Dwarfs. *Annual Review of Astronomy and Astrophysics*, 41:465, 2003

J. Isern, E. García-Berro, M. Hernanz and M. Salaris. White Dwarfs. *Contributions to Science*, 2:237, 2002

D. Koester and G. Chanmugam. Physics of White Dwarf Stars. *Reports on Progress in Physics*, 53:837, 1990

M. Salaris. White Dwarf Cosmochronology: Techniques and Uncertainties. In *The Ages of Stars*, Proceedings of the International Astronomical Union, IAU Symposium 258, 287, 2009

H. M. van Horn. *Unlocking the Secrets of White Dwarf Stars*. Springer, 2014

CHAPTER 9

Far Away Beyond the Field

In the previous chapters, I have described several methods we have devised to determine the age of individual stars and star clusters, these latter objects being populated by stars born all with the same initial chemical composition and age. The distribution of the ages of star clusters as a function of their initial metallicity, mass, and position (and orbits) within a galaxy provide a first set of constraints on the theoretical models of galaxy formation. An additional set of constraints is posed by the determination of what we call the star formation history of the field stars in a galaxy. These are simply the stars not in clusters, and make the bulk of the stellar content in any given galaxy; for example, the Sun is a field star belonging to the disk of the Milky Way. Their star formation history (SFH) is made of two pieces of information. The first one is how much gas mass has been transformed into stars at any time since the start of the galaxy formation (the star formation rate – SFR); the second piece of information is the time evolution of the metallicity of the gas, that is equal to the initial metallicity of the various generations of stars formed (the age-metallicity relation – AMR). As you can see, the SFR and AMR tell us about the evolution of the gas in a galaxy, that is difficult to predict in theoretical models of galaxy formation and evolution, as discussed in chapter 2.

The SFR and AMR are linked because each generation of stars injects in the interstellar medium – through supernova explosions and stellar winds – large quantities of gas processed by nuclear reactions, that modify the chemical composition of the remaining interstellar gas, out of which the following generation of stars form. Their determination requires to know how many stars have formed at a given time during the evolution of a galaxy (and their initial chemical composition) and this obviously entails the determination of stellar ages. It may seem a truly herculean and almost impossible task, like attempting to determine how many people have been born on Earth, let's say every 1,000 years, since the first Homo habilis appeared on this planet about 2.5 million years ago. But of course, we have an advantage when dealing with stars, for the majority of the members of each generation born since the formation of a galaxy are still 'alive', meaning that they are still shining thanks to the nuclear reactions active in their cores[1].

I describe in this chapter the main techniques to determine how many stars have formed at any given time during the evolution of a galaxy, and their initial chemical composition (the AMR). The total number of stars can then be translated to a corresponding total mass of stars (the sum of the mass of all stars formed) which is equal to the mass of the gas converted into stars (the SFR). The next section describes how to derive SFR and AMR of nearby galaxies that can be 'resolved' into their stars when observed through a telescope. For these objects, we can measure the magnitudes of individual stars and build CMDs of their stellar populations.

The final section deals with the case of more distant galaxies for which we can measure only the total ('integrated') brightness and spectrum of their stellar populations, because stars are all blended together when observed with our telescopes. It is like when we cannot make out the dot at the bottom of a question mark in a page of a book, when we stand a few metres away, and we perceive it as just one symbol. In our case the integrated brightness (or spectrum) is the sum of the brightness (or spectra) of all the stars in the galaxy within the field of view of our detector. Even when we are unable to observe individual stars, we can still derive precious information about a galaxy SFH, by making use of our knowledge of stellar evolution. The same methods that will be discussed here can, of course, be applied also to unresolved star clusters, to determine their age and initial chemical composition.

A word of caution is required. These techniques are conceptually less straightforward, and subject to larger uncertainties than the case of studying ages of individual stars or star clusters. It is an inevitable consequence of the much more complex tasks at hand. There are also

---

[1] Apart from the Milky Way, stellar remnants like white dwarfs are out of reach for even the more powerful telescopes.

several technical details and complications to deal with, which I will generally try to avoid, prioritising the basics of the methods instead.

9.1 STAR FORMATION HISTORIES OF GALAXIES

With the current generation of telescopes (both on the ground and in space) we can resolve the stellar populations and determine the SFH of galaxies out to a distance of about 1 Mpc, roughly the size of the Local Group of galaxies made of two giant spirals (our Milky Way and Andromeda) and a few dozen, smaller dwarf galaxies. The CMDs of the stellar populations of these galaxies generally reveal a more complex morphology compared to star clusters, due to the presence of multiple stellar generations. This is clearly demonstrated by Fig. 9.1, that shows the CMD of field stars of the disk of the Milky Way within 75 parsec from the Sun, with accurate distance measurements. This is a modern equivalent of Russell's CMD of the solar neighbourhood discussed in chapter 3 (see Fig. 3.3).

Notable features of this CMD are the extended (in both magnitude and colour) MS, the large number of core He-burning stars seen at M_V between 0 and 1 mag, and $(B-V)$ colours between 0.8 and 1.2 mag, and a broad SGB that departs from the MS at M_V around 4 mag, leading to a broad (in colour) RGB. The simultaneous presence of this SGB and of brighter stars still on the MS is the smoking gun pointing toward the existence of multiple stellar generations in the disk of our galaxy. The SGB marks the presence of stars that have just completed core hydrogen-burning, whose mass is smaller – hence they have a longer MS lifetime – than that of brighter objects still on the MS[2].

This is demonstrated very clearly by the composite CMD shown in the same Fig. 9.1, the combination of the CMDs of three open clusters with ages between 100 Myr and 4.5 Gyr, all with roughly the same initial chemical composition (about solar). This CMD made of just three generations of stars shows already a general qualitative resemblance with the CMD of field stars in the solar neighbourhood, displaying an extended MS, that reaches magnitudes much brighter than the SGB of the cluster M 67, about 4.5 Gyr old. Such an example shows how a basic knowledge of stellar evolution enables us to understand at first glance whether a CMD shows multiple stellar generations. However, what we

---

[2]MS stars of increasing mass are brighter and shorter lived.

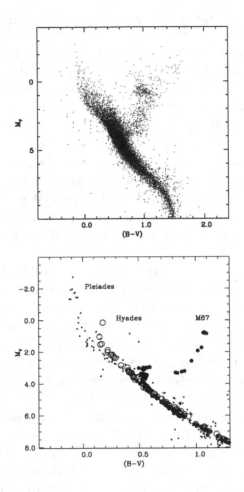

Figure 9.1 *Top panel:* CMD of stars within 75 pc from the Sun with accurate parallaxes (parallax errors below 10%) from the *Hipparcos* satellite mission [123], corrected for the effect of distance and reddening. *Bottom panel:* Composite CMD of three open clusters (the 100 Myr old Pleiades, the 900 Myr old Hyades, and the 4.5 Gyr old M 67) corrected for the individual distance moduli (the extinction for these clusters is negligible).

need is a quantitative and as accurate as possible determination of the SFR and the AMR from these complex CMDs.

As always in science (and life), we solve this problem starting from the known answers to simpler questions. We know how to determine the age of star clusters, which host only one generation of stars. A CMD with multiple stellar generations can be imagined theoretically as a sum of isochrones with different ages. If in the observed galaxy there is also an evolution of the metallicity of the stars with age, the theoretical counterpart of its CMD will be a combination of isochrones with the appropriate ages and metallicites. The question then becomes: How do we identify (in terms of age and metallicity) the various isochrones that, once combined, can reproduce the observed CMD? And also: How can we determine how many stars were born at a given age and with a given metallicity?

To address these questions, we need an additional piece of information. As discussed in chapter 4, theoretical isochrones do not only predict the CMD of a stellar population with a given age and initial composition, they also provide the value of the initial mass of the stars at a given magnitude and colour. If we know how the number of stars born in a star formation episode depend on their mass (the initial mass function – IMF), an isochrone can also predict the number of objects in a population with a given age and chemical composition. Typically, we do not need this information to determine the ages of a star cluster because all stars are coeval, but it becomes crucial when studying the SFH of galaxies.

The IMF is perhaps the most fundamental output of the star formation process, but we are still unable to predict it from first principles. However, thanks to the seminal work by Edwin Salpeter[3] in the 1950s and more recent investigations by, among others, Pavel Kroupa and Gilles Chabrier, we know from observations that the number of

---

[3] Edwin Salpeter has been a remarkable scientist. He was born in Vienna from parents both with a physics education, but emigrated with his family to Australia in 1939 at the age of 15, to escape Hitler anti-Semitic laws. He moved back to Europe to study for his PhD in theoretical physics at the University of Birmingham, in the UK, before travelling to the US to work with Hans Bethe. He has described himself as a 'generalist' – like Ernst Öpik – quipping that he learned less and less about an increasing number of subject, to reach the point of knowing nothing about everything. In addition to his work on the IMF, Salpeter has contributed to quantum-electrodynamics, nuclear physics, statistical mechanics, stellar astrophysics, extragalactic astrophysics, and also neurobiology (with his wife Mika) and epidemiology.

stars formed in a given mass range increases fast with decreasing mass [109, 69].

The IMF tells us that the ratio of the number of stars formed with two different masses M_1 and M_2 is proportional to $(M_1/M_2)^{-2.3}$ for masses larger than $0.5M_\odot$, and to $(M_1/M_2)^{-1.3}$ for lower masses. The negative sign of the exponent means that lower-mass stars are preferentially formed; for example, the number of stars formed with a mass equal to the mass of the Sun is 200 times the number of $10M_\odot$ stars. Typically, a star formation episode produces stars between 0.1 and 100 $M_\odot$, and more than 60% of them have a mass lower than the mass of the Sun[4].

With an IMF at hand, we can proceed as follows. We take a set of isochrones covering a large range of ages and initial metallicities, and for each isochrone we calculate the relative number of stars (the number ratios) point-by-point along the various branches, using the IMF described before. We can then assume that at each age and metallicity a fixed total mass of stars M_t – the same for all isochrones – has formed between 0.1 and $100M_\odot$. Once M_t is fixed, the relative numbers can be transformed to absolute numbers (actual values of how many stars are expected at a given point along the isochrone) because the sum of the masses of all the stars formed must be equal to M_t. In this way, we can calculate how many objects (we may call them synthetic stars) populate each point along an isochrone of a given age and metallicity, when a total mass of stars M_t has formed. The mass range spanned by the MS portion of an isochrone is much larger than the following evolutionary phases (see chapter 4) hence the MS will be much more populated than more advanced phases. This is just a consequence of the longer evolutionary times during core hydrogen-burning[5].

After this step, we add together the synthetic stars along each

---

[4]The IMF is generally assumed to be universal, but there is a long-standing debate on whether the exponents (the quantities -2.3 and -1.3) vary for example between different types of galaxies.

[5]In chapter 5 I have calculated the number of stars along the RGB of an isochrone using the approximation that this phase corresponds to the RGB of an evolutionary track with an initial mass equal to the isochrone turn-off mass. In that case, the relative number of stars along the RGB was a function of the speed with which the corresponding model evolves. That procedure gives exactly the same result as using the IMF and the very narrow range of masses along the isochrone RGB. The reason is that, as discussed in chapter 4, the range of masses across any branch of an isochrone is fixed by the evolutionary speed of the models in the corresponding evolutionary phase.

isochrone, eventually rescaling the predicted numbers of objects by varying M_t (upwards or downwards) for individual isochrones, until we find the optimal combination of ages t, metallicities Z and the corresponding total mass of stars formed M_t (that can vary with t and Z) such that the morphology and number of stars across the observed CMD is best matched. Complex statistical algorithms are employed to achieve this best match, and the resulting isochrone t, Z, and M_t provide the SFH of the observed galaxy. We can find a unique combination of t, Z, and M_t from this procedure because colours and magnitudes of different parts of an isochrone have different sensitivities to metallicity and age (see chapter 4 and chapter 7). In addition, the evolutionary timescales, hence the mass distribution and number of stars along different branches of an isochrone for a fixed M_t, do depend on the initial metallicity of the stellar population. These are the properties that can disentangle the effects of age, initial composition and M_t when we fit an ensemble of isochrones to the CMD of a galaxy. And depending on the apparent size of the target galaxy and the field of view of our detector, we can even determine with this technique the SFH in different areas, to study its variations (if any) across the galaxy.

In the case of some nearby dwarf galaxies, there are also spectroscopic chemical abundance determinations available for bright RGB stars (like for the Sculptor dwarf galaxy [29]). In this case, the synthetic stars that match the observed CMD have to match also the distribution of the observed RGB abundances.

This is the essence of the method, pioneered by groups in Bologna and Padua [36, 14], later expanded and refined by multiple researchers [2, 30, 32, 44, 53, 116, 122]. It is like the example in the composite CMD of Fig. 9.1, adding together CMDs of open clusters of different ages, to reproduce the morphology of the CMD of solar neighbourhood stars. But to do things properly, we need to match also the star counts across the whole observed CMD, not just the CMD morphology. The age and metallicity distributions we retrieve this way match those of the target stars in a statistical sense; we do not determine ages and metallicity on a star-by-star basis, rather their global distributions.

Another way to describe the idea underpinning this method is to think of colours. There are three primary colours, namely red, yellow and blue, plus black and white. Any other colour can be made by mixing these five ones. For example, orange is a mixture of yellow and red, and brown is a mixture of red and green. Adding white makes the colour lighter, and adding black makes it darker. Finding the SFH of a stellar population from its CMD is like finding the proportions of yellow, red, white, and black (isochrones of different ages and metallicities) that have produced the orange colour (the CMD of the target stellar population) of a backpack seen in a shop.

Typically, the HB of old populations is excluded by the match with isochrones, because we are unable to predict the amount of mass lost by RGB stars[6]. Distance modulus and reddening of the target galaxy are usually assumed beforehand, from independent determinations. These techniques, however, can in principle also estimate distance and reddening, like when performing isochrone fitting on a star cluster (see chapter 7).

It is also essential that the observed CMD reaches magnitudes below the oldest MS turn-off, to be sure we determine the age of the oldest stellar generations, and find the complete SFH. This limits the distance of the galaxies for which we can determine the full SFH from their CMDs.

There is however a complication, related to the unavoidable measurement errors. Stars belonging to a population with a given Z and age won't lie on a narrow sequence in the CMD, as the corresponding isochrone, because of the errors on the measured magnitudes (the 'photometric error'), like the case of the CMDs of star clusters briefly mentioned in footnote 21 of chapter 7. Besides, unresolved binary stars (we can measure only the sum of the brightnesses of the two components) also cause a broadening of the MS, beyond the effect of initial chemical composition and age. We can take into account these effects statistically, by adding to the magnitude and colour of each synthetic star populating an isochrone the random effect of a photometric error (the value of photometric errors as a function of magnitude can be derived from the analysis of the observations) and the systematic effect of unresolved binaries, making assumptions about the distribution of the masses – hence position in the CMD – of the unresolved companion, and the value of the binary fraction[7]. These synthetic stars will spread in the CMD around the narrow sequence of their parent isochrone, and as a consequence the CMDs produced by isochrones

---

[6] Alessandro Savino, Eline Tolstoy, and I have recently devised a technique that, under certain assumptions, can derive both the SFH and RGB mass loss in galaxies with old stellar populations, including also HB stars in the CMD matching [116].

[7] All of this can be done fairly easily with computers, using appropriate numerical routines and random number generators that simulate the probabilistic nature of measurement errors and values of the masses of binary companions.

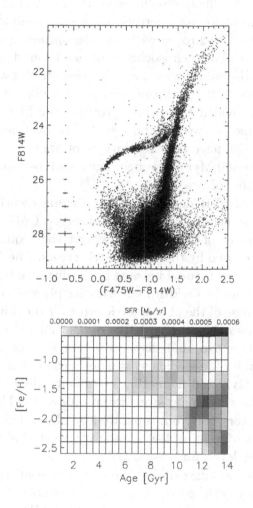

Figure 9.2 *Top panel:* CMD of the Tucana dwarf galaxy from Hubble Space Telescope observations. The apparent magnitude on the vertical axis corresponds roughly to the I filter, and the colour on the horizontal axis to $(B - I)$ [79]. *Bottom panel:* Derived SFH for the Tucana galaxy [116]. Each box corresponds to a combination of age (in Gyr) and initial metallicity (given as [Fe/H]). They greyscale at the top denotes the strength of the star formation, given as solar masses of gas transformed into stars per year (courtesy of Alessandro Savino).

with too small age and/or metallicity differences may overlap, limiting the age and metallicity resolution of the derived SFH. Depending on the size of the photometric errors, we need to employ in the CMD matching a grid of isochrones with an appropriate spacing in age δt and metallicity δZ. For each isochrone of age t found to contribute to the observed CMD, we then assume it represents the whole age range up to the next age in the grid; the derived total mass M_t of stars born at age t is then divided by δt, and provides the SFR (mass of stars born per unit time) between t and $t + \delta t$. The best determinations of the SFH of galaxies have time resolutions of about 1–2 Gyr for ages above 1 Gyr, 200–400 Myr at younger ages, and metallicity resolutions typically of a factor of two.

An example of SFH derived for the Tucana dwarf galaxy in the Local Group, is shown in Fig. 9.2. The observed CMD resembles that of a globular cluster of the Milky Way, but the sequences are wider than the spread caused by the photometric errors. The MS does not go much deeper than the turn-off because of the distance (equal to approximately 0.9 Mpc) and its width increases sharply with magnitude, due to the rapid increase of the photometric error for the faintest stars. The derived SFH shows that the majority of Tucana's stars formed more than about 11 Gyr ago, and displays a broad range of metallicities. There is also a tail of less efficient star formation that reaches ages as low as about 5 Gyr, and increasing metallicity. The SFH appears to comprise three distinct main star formation events. The two stronger events occurred very early on, separated by about 1–2 Gyr, while the last star formation event, of lower intensity, started about 10 Gyr ago and lasted at low levels for several Gyr.

Shouldn't we be amazed that we can derive all this information about the history of a galaxy from such a simple diagram – just a brightness and a colour of its stars – like this CMD?

9.2   UNRESOLVED STELLAR POPULATIONS

As mentioned at the beginning of this chapter, for the overwhelming majority of the galaxies (and star clusters) observed with our telescopes, we can measure only the integrated brightness and integrated spectrum of their stellar populations, because stars appear all blended together. When we observe an unresolved galaxy (or star cluster) – or an area within an unresolved galaxy, depending on its size and brightness – for example in two photometric filters, instead of a CMD like V

versus $(B - V)$ containing tens of thousands of stars, we can measure only one value of V and one value of the colour $(B - V)$. Most of the information encoded in the magnitudes of the individual stars appears to be lost, and it may sound truly insane to even think of determining the ages of unresolved stellar populations.

But maybe it isn't as insane as it seems. As early as 1957 William Morgan and Nicholas Mayall published integrated spectra of a sample of galaxies, noticing at the beginning of their study that the observed absorption and emission features, despite their composite nature, must contain significant information about the stellar populations of the systems [80]. They even attempted to infer the shape of the CMD of the stellar populations of a few galaxies (Andromeda being one of them) based on the spectral features observed, a sort of qualitative estimate of the galaxies' SFH. In 1961 Joan Crampin and Fred Hoyle showed that the integrated colours of unresolved stellar populations vary with age, and speculated about determining the age of galaxies from their colours [28], while in 1966 David Wood tried to determine the population ratios among stars in different parts of the CMD of several galaxies, using as constraints integrated colours and some features of their integrated spectra [127].

A couple of years later, the seminal study by Beatrice Tinsley – actually her PhD thesis – finally paved the way for the quantitative interpretation of integrated photometry and spectroscopy of galaxies [121].[8]

9.2.1 Integrated magnitudes

Let's stick with photometry for the time being: The first step to see if we can learn something about the age of unresolved systems is to calculate the predicted integrated magnitudes of simple stellar populations like star clusters, which host stars all with the same age and initial

---

[8]Despite a short career (she died at 40) Beatrice Tinsley work has set the foundation for the modern studies of galaxy evolution, her PhD thesis being described as 'extraordinary'. She was born in Chester, an ancient Roman town in the North-West of England, not far from Liverpool, but grew up in New Zealand and then moved to the US with her then-husband Brian Tinsley. Her career spanned just 14 years starting with the PhD thesis of 1967, followed by about 100 scientific papers. The New Zealand Geographic Board has named a mountain in her honour, Mt Tinsley in the Kepler Mountains of Fiordland. Asteroid 3087 discovered in 1981 is named Beatrice Tinsley.

chemical composition. To do so, we need to pick an isochrone[9], fix the total initial mass M_t of our fictitious unresolved stellar population, calculate the number of stars at each point along the isochrone using the IMF (as described in the previous section), and add the brightness in the chosen photometric filters of the synthetic stars along the whole isochrone[10].

When we calculate integrated magnitudes from isochrones with a range of ages and initial metallicities, we learn something very interesting. First of all, obviously the larger M_t the brighter the integrated magnitudes for any given age t and Z: This is simply a consequence of having more stars in our fictitious stellar population when M_t is larger. At fixed initial chemical composition and M_t, the integrated magnitudes on average tend to increase with increasing age, meaning that they get fainter. The reason is that the maximum brightness of the isochrones and the brightest point along the MS tend to become fainter with increasing age (see the isochrones shown in chapter 4). Also a change of metallicity affects the integrated magnitudes at fixed age t and M_t, making them generally fainter with increasing Z.

This behaviour, however, doesn't help us to derive the age of our target populations, because we generally do not know its mass, hence individual integrated magnitudes are not useful for our purposes. Moreover, changes of initial composition and age both affect the integrated magnitudes.

We can hope that integrated magnitudes in distinct filters fade at different speeds when the age increases, making integrated colours (the difference of integrated magnitudes in two filters) age dependent. Colours have the advantage to be unaffected by the mass of the population[11] and they are also independent of the distance to the target.

---

[9]The isochrone does not need to include also the WD sequence, because WDs give a negligible contribution to the integrated magnitudes. The HB phase in old stellar populations needs, however, to be included, because HB stars are bright and relatively long-lived. This introduces an intrinsic uncertainty in the predicted integrated magnitudes of old populations, especially at the low metallicities typical of globular clusters, because of our inability to predict the HB morphology for these populations.

[10]The brightness is the quantity $10^{-0.4M_A}$, where M_A is the magnitude of the synthetic star in the generic filter A. The final integrated magnitude is -2.5 times the logarithm of the sum of the brightness of all synthetic stars.

[11]We have seen in chapter 6 that colours depend on the ratio of the brightness in the two filters; in case of integrated colours, the ratio of the sum of the brightness of the synthetic stars along the same isochrone depends only on their relative numbers along the isochrone.

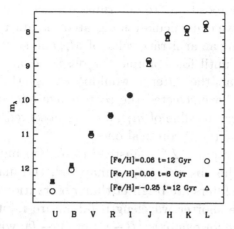

Figure 9.3 Integrated magnitudes in several photometric filters (from left to right in order of increasing wavelength) for three fictitious populations with the labelled ages and [Fe/H] values. The sets of magnitudes of the three populations are shifted to the same value in the I filter.

It turns out that the speed of change of the integrated magnitudes with age does indeed vary from one filter to another. Different photometric filters cover different wavelength ranges, therefore the brightness of the same stars varies from one filter to another because different parts of their blackbody spectrum are sampled. For example, the integrated light sampled by an infrared filter like K is dominated by bright, cool stars, like RGB and AGB stars, whilst a filter like B samples mainly the contribution of hotter MS stars. Unfortunately, the effect of a change of age on the colours can be mimicked by changes of composition at fixed age. An increase of the age at fixed metallicity generally makes colours redder, but also an increase of metallicity at a fixed age has the same qualitative effect.

Figure 9.3 shows the integrated magnitudes of three fictitious populations in several photometric filters; the ensemble of filters cover a very broad wavelength range, from the near-ultraviolet around 350 nm for the U filter, to the infrared around 3500 nm for the filter L. Let's consider as reference the 6 Gyr old population with [Fe/H]=0.06 (metallicity 15% above the solar value), whose absolute magnitudes (calculated for a given value of M_t) have been shifted vertically to account for an arbitrary distance. We take these values as an observed set of apparent integrated magnitudes of a stellar population with a single age and initial chemical composition. To determine its age and metallicity (assumed unknown) we can calculate integrated magnitudes from various isochrones assuming an arbitrary value of M_t, and shift them vertically in this diagram[12] until for example the observed m_I is matched (we could have picked another filter, it wouldn't change the outcome of this discussion). If we have selected the correct age and metallicity, after matching the observed value of m_I all other observed magnitudes will also be matched by the theoretical ones.

If the age is too old (12 Gyr instead of 6 Gyr) but the metallicity is correct, when the observed m_I is matched, the magnitudes in the longer wavelength filters are brighter than observations, conversely the magnitudes in the shorter wavelength filters are fainter. This means that all colours like for example $(B - V)$ or $(V - I)$, with the first filter at shorter wavelength than the second one, are larger (redder) than observations. If I calculate integrated magnitudes for a metallicity a factor of two lower ([Fe/H]$=-0.25$) after the observed m_I is matched, all observed integrated magnitudes from U to I are also almost exactly matched, for an age that is twice the real age of the target population. All colours involving filters from U to I therefore suffer from what is called the 'age-metallicity degeneracy', meaning that observed colours in this wavelength range can be matched by different combinations of age and metallicity. As we can see from this example, a metallicity too low in the models can be compensated for by an age too high.

If we also consider infrared filters, the situation improves; the lower metallicity-higher age combination fails to match the integrated magnitudes of filters at wavelengths longer than I. This suggests using pairs of colours that include at least one infrared filter, to determine both age and initial metallicity of our population.

In practice, we can employ integrated colour-colour diagrams like the one in Fig. 9.4[13]. The predicted colour sequences of constant age at fixed metallicity, and constant metallicity at fixed age are roughly orthogonal (they cross at an angle generally not too far from 90 degrees):

---

[12]The effect of the distance is just a vertical shift, the same for all magnitudes, in the assumption that extinction has been corrected for, or is zero. The same is true for the effect of the unknown M_t of the target population: A variation of M_t changes all integrated magnitudes by the same amount.

[13]This diagram, which I have developed together with colleagues, is one possible choice [63]. There are several other possible colour combinations that break the age-metallicity degeneracy, like for example the $(V - I)$-$(V - K)$, $(U - R)$-$(R - K)$, or $(B - R)$-$(R - K)$ diagrams [12, 86, 91].

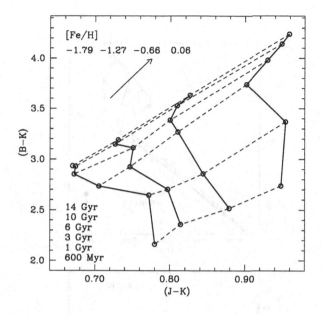

Figure 9.4 Example of theoretical integrated colour-colour diagram that breaks the age-metallicity degeneracy [63]. Solid lines are sequences of constant [Fe/H] and varying age, whilst dashed lines are sequences of constant age and varying metallicity. Age increases towards the top of the diagram, [Fe/H] increases towards the right. The arrow displays how observations are displaced by the effect of extinction (both colours become redder, and move in this diagram following the direction of the arrow).

If we put in this diagram a point corresponding to observations of an unresolved stellar population, we can then easily determine its age and metallicity from the position of the measured colours relative to the lines of the theoretical calibration. A common feature of all these diagrams is that the spacing between lines of constant age gets narrower at older ages, especially at low metallicities. It is very hard, for example, to discriminate between a 10 and a 14 Gyr old population, given that typical measurement errors are of several hundredths of a magnitude, comparable or larger than the spacing between the 10 and 14 Gyr sequences in Fig. 9.4.

This diagram stops at a minimum age of 600 Myr, and the following Fig. 9.5 shows what happens at younger ages. The upper panel of

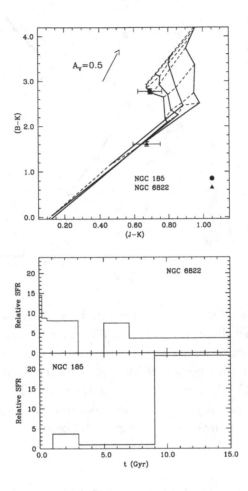

Figure 9.5 *Top panel:* The same colour-colour diagram of Fig. 9.4 extended to ages below 600 Myr. Ages of 300 Myr and 40 Myr have been added to the diagram. At 300 Myr the lines of constant age and metallicity are still reasonably orthogonal, but between 300 and 40 Myr lines of constant age and metallicity almost overlap, making impossible to disentangle these two quantities. The effect of an extinction A_V=0.5 is also shown, together with the observed integrated colours – corrected for extinction – of two galaxies. *Bottom panel:* Time evolution of the SFR for the galaxies in the top panel, as determined from their CMDs. The relative strength of the SFR at different ages is displayed.

this figure tells us that at 300 Myr the age-metallicity degeneracy is still reasonably well broken, but between 300 Myr and 40 Myr the sequences of constant age and metallicity basically overlap, precluding the derivation of both quantities from this diagram. In general, this type of integrated colour-colour diagrams can break the age-metallicity degeneracy only for ages not much lower than 1 Gyr.

Another problem we have to face is that in the case of galaxies with multiple generations of stars, these diagrams provide us with one single pair of age and metallicity values, which depend on the global distribution of these two quantities. To highlight this point, the same Fig. 9.5 shows an interesting experiment, involving the two Local Group galaxies NGC 185 and NGC 6822, whose integrated colours are displayed in the upper panel. These relatively nearby galaxies can also be resolved into their individual stars with sufficiently powerful instruments, like the Hubble Space Telescope. In this case, we have the luxury to be able to derive the SFH as described in the previous section. The lower panel of the figure displays the SFR obtained from the CMDs of the stars of these two galaxies (I focus here just on the ages): NGC 185 had an active SFR down to ages around 9 Gyr, followed by a low-level star formation until 1 Gyr ago, whilst NGC 6822 has formed stars more or less continuously since right after the Big-Bang, with an increased rate in the last few billion years. The colour-colour diagram provides an age around 3 Gyr for NGC 185, and a much younger – below 300 Myr – but undefined age for NGC6822, because its colours are in the region of the age-metallicity degeneracy.

Despite NGC 185 has formed most of its stars at ages above 9 Gyr, the single age derived from the diagram is much younger than this value. This is a general result for all colour-colour diagrams: Even small fractions of young stars bias the derived single age towards low values. The reason is that young populations are generally much brighter than old ones and always tend to dominate the integrated magnitudes of multiple generation systems. The very young – but undetermined – age obtained for NGC 6822 is also a consequence of this general property[14].

Comparisons among ages of unresolved galaxies obtained from colour-colour diagrams are therefore of difficult interpretation, and

---

[14] An obvious question then springs to mind. If, for example in case of NGC6822, I take the SFH determined from the CMD and calculate the corresponding integrated colours, where do they fall in the diagram of Fig. 9.5? Well, that's what I have checked and, pleasantly enough, the theoretical colours match the observed ones [63].

cannot disentangle the effect of a complex SFH. An alternative possibility is to assume a mathematical form for the SFR of a galaxy, or a class of galaxies (obtained for example from models of galaxy formation or other considerations) and use observed colours and colour-colour diagrams to fix the free parameters of this SFR [118]. As a simple example, we might have reasons to think that our target galaxy had a constant SFR – for simplicity let's assume that also the metallicity is constant with age, but unknown – starting t_i years ago (t_i is the unknown age of the oldest stars in this hypothetical galaxy), and use a colour-colour diagram to determine both t_i and the metallicity of the galaxy. To this purpose we would calculate integrated colours for this type of SFR and various pairs of t_i and Z, to find the values that allow us to match the observed colours.

9.2.2 Integrated Spectra

The study of integrated spectra of unresolved stellar populations provides us with other powerful diagnostics of their age and SFH. Integrated spectra include a large number of spectral features sensitive to the surface chemical composition, T_{eff} and surface gravity of the stars; if we can decode this information, we might be in the position to at least constrain ages and SFH of our target population, without the need of any *a-priori* assumptions.

As for stars, measurements of integrated spectra are obviously more challenging than those of integrated magnitudes. At the distances of extragalactic systems, we cannot measure spectra with the same level of details of nearby individual stars, meaning that we need to add together photons within wavelength ranges on the order of, let's say 0.1 nm, instead 0.01–0.001 nm. Moreover, due to the motions of the stars within the host galaxy, the combined effect of the Doppler shifts affecting each individual stellar spectrum broaden the various features in the measured integrated spectrum (they cover a wider wavelength range compared to the case of no stellar motions) because stars are both moving towards us and away from us along the line of sight.

Figure 9.6 displays two examples of spectra of galaxies (one elliptical and one spiral) at optical wavelengths. The spectrum of the elliptical galaxy shows only absorption features, whilst the spiral galaxy counterpart also contains emission features, like the one around 660 nm. The absorption features[15] are caused by the absorption lines produced in

---

[15]Notice that I call them 'features' and not 'lines', because they are actually the

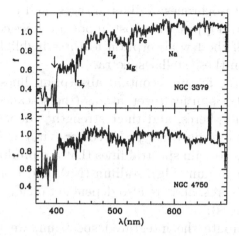

λ(nm)

Figure 9.6 Integrated spectra at optical wavelengths of an elliptical galaxy (upper panel) and a spiral galaxy (lower panel) with the brightness rescaled to have a value equal to unity at 550 nm [64]. Some important absorption features in the spectrum of the elliptical galaxy are labelled, together with the 'break' at 400 nm (marked with an arrow).

the atmospheres of the unresolved stars, while emission features are explained by the emission spectrum of low-density interstellar gas heated by nearby hot and young stars.

The quantitative interpretation of absorption features in integrated galaxy spectra (but also in star cluster spectra) is more complex than the case of stars. To calculate the integrated spectrum of a stellar population of a given age and chemical composition we need to pick the corresponding isochrone, assign an appropriate spectrum to each point along the isochrone, fix the total initial mass M_t, and calculate the number of synthetic stars at each point using the IMF. Finally, for each wavelength, we add the brightness of the spectra of all these synthetic stars[16].

To highlight the properties of these integrated spectra let's start by looking at the top panel of Fig. 9.7, showing stellar spectra of solar composition representative of spectral types A F G K. The lower panel

---

blend of more than one spectral line, due to the resolution of the spectra of these distant objects.

[16]When comparing with observations, we also need to 'simulate' the properties of the observed spectrum in terms of resolution, and broadening due to the motion of the stars in the target galaxy.

shows the HRD of isochrones of different ages, and the T_{eff} range for each of these spectral types. The most important lines corresponding to the wavelengths of the few absorption features highlighted in Fig. 9.6 are also marked on these stellar spectra.

The H_γ and H_β features contain absorption lines corresponding to electron transitions in hydrogen atoms (the so-called Balmer lines) in the stellar atmospheres, and their strength[17] depends on T_{eff}: It increases moving from spectral type K (cooler stars) to A (hotter stars). The other features contain spectral lines that are produced by electron transitions in magnesium (Mg), sodium (Na), and iron (Fe) atoms in the stellar atmospheres, and are also dependent on T_{eff} (and of course chemical abundances).

When we compute the integrated spectrum, we have to add at each wavelength the contribution of the spectra of all points along the isochrone. Focusing for example on the wavelength regions of the two Balmer lines, hot and bright A stars in the MS will contribute relatively low photon numbers in the H_γ and H_β regions, compared to the neighbouring wavelengths (H_γ and H_β lines are very strong in these stars); on the other hand, K stars will contribute a comparable number of photons in the H_γ and H_β ranges and at the neighbouring wavelengths. As a result, the strength of the H_γ and H_β features in the final integrated spectrum depends on the T_{eff} (spectral type), luminosity (brighter stars give off more photons), and number distributions of the various types of stars within the population, even stars that do not show the line in their spectra.

Luminosities, effective temperatures, and number distribution of stars in a population depend on their ages and chemical compositions, their SFH[18]. Similar conclusions can be drawn regarding, for example, the Na feature (and all other features in the integrated spectra); The corresponding Na spectral lines are strong in cool K stars and much weaker or non-existent in hotter stars. The strength of the feature in the integrated spectrum depends therefore on the Na abundance, which affects the line strength in K stars, but also on the distribution in the HRD and the relative numbers of stars of all spectral types.

Another important feature of integrated spectra of stellar populations, marked with an arrow in Figs. 9.6 and 9.7, is the so-called

---

[17]The strength of spectral lines and absorption features is measured by the area they cover in a brightness-wavelength diagram.

[18]For coeval populations the strength of H_γ and H_β increases with decreasing age.

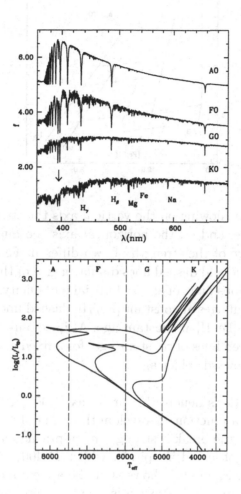

Figure 9.7 *Upper panel:* Optical spectra of stars representative of the labelled spectral types. The position of the same spectral features of Fig. 9.6 and the 'break' at 400 nm are marked. The vertical scale displays for each spectrum a relative brightness, and the various spectra are shifted vertically by different amounts. *Lower panel:* HRD of isochrones for 800 Myr, 1.5 Gyr, 10 Gyr, and initial solar chemical composition. The T_{eff} ranges corresponding to the spectral types A F G K are also marked.

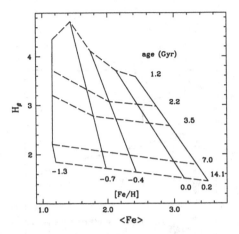

Figure 9.8 Diagram showing on the vertical axis the strength of the H_β absorption feature, and on the horizontal axis the quantity $< Fe >$, that is the average of the strength of two different Fe absorption features [42]. The solid and dashed lines display a grid of theoretical values calculated for various pairs of age and initial metallicity. Solid lines correspond to different ages at constant [Fe/H], dashed lines correspond to different values of [Fe/H] at constant age. As in colour-colour diagrams, the spacing between lines of constant age, for a fixed age difference, becomes narrower towards older ages.

break at 400 nm (let's denote this break as $D(400)$), a sharp jump of the brightness as a function of wavelength around 400 nm. The size of $D(400)$ is larger – the break is strong – in elliptical galaxies, compared to weaker breaks observed in spirals, whilst $D(400)$ is almost zero in irregular galaxies. Figure 9.7 shows that the appearance or disappearance of the break in galaxy spectra is due to a combination of effects. The spectrum of the K star shows the jump at 400 nm caused by strong absorption at shorter wavelengths, while the spectrum of the much hotter A star displays no break, and actually its brightness tends to peak around those wavelengths. The value of $D(400)$ is therefore strongly dependent on the number fraction of young, hot MS stars in the target population, whose presence tends to suppress the jump at 400 nm in the integrated spectrum. As a consequence, the measured $D(400)$ is sensitive to the SFH of a galaxy and can be used as an indicator of some sort of mean age (that depends on the details of the SFH) of its stellar populations.

These few examples should be sufficient to demonstrate that the integrated spectrum of a galaxy (or a star cluster) contains a wealth of information about the chemical composition and ages of the underlying stellar population. Diagrams like the one in Fig. 9.8 have been extensively used to study the ages of unresolved populations, and are the counterpart of the colour-colour diagrams discussed previously, for lines of constant age and constant initial metallicity are roughly orthogonal. The measured values of H_β and $< Fe >$ (the average of the strength of two Fe absorption features in the integrated spectrum) are sufficient to determine the age and [Fe/H] of the target population from a comparison with theoretical grids like the one shown in the figure[19]. As with colour-colour diagrams, it is impossible to disentangle the multiple stellar generations in case of a complex SFH. This diagram provides us with just one pair of age and [Fe/H] values and in the case of several stellar generations, the derived age is always heavily biased towards the age of the youngest populations, whilst the derived metallicity is generally biased towards that of the oldest component.

To exploit as much as possible all information encoded in the observed integrated spectra, and determine the full SFH of an unresolved galaxy, we can take advantage of the power of modern computers and statistical algorithms, to perform a 'full-spectrum fitting' [26, 68, 82, 84, 126]. The method is analogous to finding the SFH of a resolved galaxy from its CMD: By varying the values of the initial total mass M_t, we search for the best combination of theoretical integrated spectra calculated for single-age, single-metallicity populations that match the observed spectrum – the brightness at each wavelength, including all absorption features – of the target galaxy[20]. The result of this matching provides age, metallicity and initial mass M_t of the various generations of stars that contribute to the galaxy stellar population. To obtain the correct values of M_t the distance to the galaxy must be known[21]; another parameter to be fixed (but can also be derived from the fitting procedure) is the extinction, which tends to change the

---

[19]The index strength does not depend on the total mass of the stars, like for integrated colours. It is the relative brightness of the spectrum as a function of the wavelength that determines shape and area of the absorption feature.

[20]The fitting procedure excludes emission lines – if present – because they are produced by the interstellar gas.

[21]The value of the initial total mass M_t provides a scale factor of the brightness at the various wavelengths, the same for every λ. The same is true for the distance, that needs to be known to fix the values of M_t.

overall slope of the brightness versus wavelength, due to its dependence on the wavelength briefly discussed in chapter 6.

This full-spectrum fitting is performed usually on spectra in the optical wavelength range, like the examples in Fig. 9.6, and the time resolution of the SFH is typically more coarse that in the case of resolved galaxies. In worst-case scenarios, due to the faintness of the targets and the large uncertainty on the measured spectrum, we can just determine the SFH for ages above let's say 0.5–1 Gyr, and below this threshold. With better data, the time intervals δt of the SFH are narrower and have often a constant spacing on a logarithmic scale. For example, a δt equal to 0.2 dex in logarithmic scale mean that the ratio of the ages at the boundaries of each interval is equal to 1.58 (the result of $10^{0.2}$): In this case, the oldest age interval of the SFH is taken for example between 14 and 8.8 Gyr, the next one between 8.8 Gyr and 5.6 Gyr, and so on. This way we take into account the fact that integrated spectra (and this is also true, to various degrees, for integrated colours and isochrones) are less and less sensitive to a change of age (it is harder to tell them apart) when this latter increases.

If the galaxy is distant enough for the cosmological redshift z to be non-negligible, a correction needs to be applied to the observed spectrum, for subtracting the effect of the wavelength shift of the photons before comparing with theory[22]. Moreover, when we deal with these objects, we are seeing the galaxies as they appeared substantially earlier in the evolution of the universe: For example at $z=0.1$, the universe was about 1.3 Gyr younger than today. This is the lookback time already discussed in chapter 2, which at low-redshift is negligible compared to the resolving power of our age and SFH determinations.

The increase of lookback time with z can be used to our advantage when studying galaxy formation and evolution. We can in fact directly see how galaxies appeared in the past at different stages during their evolution, and determine what I would call effective ages – that depend on the SFH – at that time, for example from integrated colour-colour diagrams or the $D(400)$ break, or even some coarse SFH if data allows (these effective ages and SFHs of course need to take into account that for these galaxies the age of the universe was appreciably lower than today).

It is truly breathtaking to consider that we now can, to a certain extent, unveil the age and chemical composition makeup of stars

[22] An analogous correction obviously needs to be also applied if we work with integrated colours.

in galaxies so far away, that they appear blended together in distant blobs of light, even with the most powerful telescopes: We just need to measure their combined spectrum and use our well-established theory of stellar evolution. I cannot but wonder how William Wollaston and Joseph Fraunhofer, who over 200 years ago first observed spectral lines in the Sun, would react to the knowledge of what their observations have led to.

GLOSSARY

Age-metallicity degeneracy: Degeneracy between the effect of varying metallicity at a constant age, and varying age at a constant metallicity, on the integrated magnitudes and colours of a stellar population.

Age-metallicity relation: Time-evolution of the initial metallicity of stars formed in a stellar system.

Initial mass function: Empirical function that describes the relative numbers of stars of different mass born in a star formation event.

Integrated magnitude: Magnitude obtained from the sum of the brightness of all the stars in a stellar system.

Integrated spectrum: Spectrum obtained from the sum of the spectra of all the stars in a stellar system.

Star formation rate: Time evolution of the total mass of gas transformed into stars in a stellar system.

FURTHER READING

M. Cignoni and M. Tosi. Star Formation Histories of Dwarf Galaxies from the Colour-Magnitude Diagrams of Their Resolved Stellar Populations. *Advances in Astronomy*, 158568, 2010

C. Conroy. Modeling the Panchromatic Spectral Energy Distributions of Galaxies. *Annual Review of Astronomy and Astrophysics*, 51:393, 2013

A. Renzini and L. Greggio. *Stellar Populations*. Wiley-VCH, 2011

What Now?

In a 1927 lecture organized by the Carnegie Institutions of Washington, Edwin Hubble said that the history of astronomy is the history of *receding horizons*, first restricted to our solar system, and eventually extended to the Milky Way and further on to the extragalactic universe, once we discovered that there are other galaxies beyond ours. The trajectory of the scientific determination of the ages of stars has followed a similar course; it has progressed from the Sun to reach nearby stars and star clusters in our galaxy, then moved out to other galaxies and extragalactic star clusters, reaching systems so remote that individual stars cannot even be resolved with our detectors.

Thanks to the methods discussed in the previous chapters, we know that the stars we observe across the various galaxies in the universe do not have all the same age, as well as there isn't a 'universal' distribution of stellar ages common to all galaxies. Whilst the Sun is about 4.5 Gyr old, the oldest stars in our galaxy reach ages around 13–13.5 Gyr, about the age of the universe as determined from the current cosmological model. Stars in the halo appear to have formed almost exclusively within the first 2–3 Gyr after the Big-Bang and the bulge contains mainly a population as old as the halo, likely together with a fraction of much younger stars. New generations of stars have instead been born continuously – but not with a uniform star formation rate – in the disk of the Milky Way since the beginning of its formation, which started at about the same time as the halo. The detailed behaviour of the SFR and AMR of disk stars (not only of the Milky Way) depend however on their position within the disk and the properties of their orbits [1, 98]. The range of ages of the open and globular clusters of the Milky Way mirror that of the field disk and halo stars, respectively, and few open clusters have ages comparable to those of globular clusters.

Other spiral galaxies have formed stars more or less continuously since the Big-Bang, qualitatively like the Milky Way, and generally, but for the lowest mass spiral galaxies, the main star formation activity in the bulge stopped at old ages, like in the Milky Way. There is however a fraction of spirals (on the order of tens of percent, increasing with galaxy mass) that despite some ongoing star formation appear to be populated mainly by old stars, with almost all the stellar mass formed before a few billion years ago ('red spirals'). Elliptical and lenticular galaxies have formed stars preferentially at old ages: Over 90% of the stars in massive elliptical galaxies are older than about 10 Gyr, a fraction that however decreases with decreasing galaxy mass.

We have also been able to determine the detailed age distribution of stars in almost all dwarf galaxies in the Local Group, discovering a large variety of star formation histories. While all these galaxies host stars older than 10 Gyr, the time evolution of the star formation rate varies wildly from one object to another. For example, the Carina dwarf galaxy had well separated and short-lived bursts of star formation about 12, 9, 5, and 2 Gyr ago, while the galaxy Sculptor experienced a prolonged star formation starting about 13-13.5 Gyr ago, slowly decreasing to zero after about 8 Gyr, and the Aquarius dwarf galaxy had a roughly constant star formation until today. Again, the details of the SFR and AMR seem to depend on the position within the galaxy.

This is just a very broad-brush summary of some main results about stellar populations in galaxies of the local universe, representative of the kind of information gathered from the determination of stellar ages, that is fed into theoretical models of galaxy formation and evolution. Regarding the age of the universe, the more precise estimates of the ages of the oldest stars are still those of the oldest Milky Way globular clusters, that give consistently values around 13 Gyr with an uncertainty of about 1 Gyr around this age.

Moving to objects closer to human experience, after the discovery of the first planet outside the Solar System ('exoplanet') in 1992, we have now found (at the time of writing) about 4350 exoplanets and this number is steadily increasing. At the beginning of 2020 also came the news of the discovery of a planet approximately the same size of Earth, in the so-called habitable zone – the range of distances from the host star, in which a rocky planet like Earth can have liquid water on its surface, hence possibly support life – around a main sequence star with both mass and radius smaller than the Sun, at a distance of about 100 light-years.

With such a sizable sample of exoplanets, we can now start investigating whether the properties of planetary systems depend somehow on the age of the host stars, with the aim of testing and constraining the theoretical models of planet formation and evolution. The most common exoplanets discovered to date have radii between the Earth radius (denoted with $R_\oplus$) and the radius of Neptune, about four times $R_\oplus$. Objects with radii between 1 and 1.6–1.8 $R_\oplus$ should be rocky, with 'Earth-like' densities, the 'super-Earths', whilst planets with radii above this threshold up to 4 $R_\oplus$ have lower densities, suggesting thick hydrogen and helium envelopes around a solid core – similar to Uranus and Neptune in the solar systems – the 'sub-Neptunes'. A first recent result from the determination of the ages of planet-host stars (that span a range from a few hundred million years up to more than 10 Gyr, albeit with generally large uncertainties) shows that the distribution of the radii of exoplanetary systems seem to depend on the age of the central star [13]: With increasing age the fraction of super-Earths to sub-Neptunes increases. This result, if confirmed with an even larger sample of exoplanets and more accurate ages, poses important constraints on the formation and evolution of planetary systems.

10.1 HOW ACCURATE ARE OUR AGES?

We have seen in chapter 3 how the question of the age of stars could finally be addressed scientifically in the nineteenth century, taking advantage of the newly discovered laws of thermodynamics. These first studies gave ages on the order of tens of millions of years, which in the following century steadily increased to over 10 Gyr for the oldest stars in the universe. In their book *Science Askew*, Donald Simanek and John Holden notice, somewhat tongue-in-cheek, that the rapidly increasing value of the age of the universe in the last couple of centuries (from millions of years to more than 10 Gyr) will lead in the future to an infinite value; as a consequence, the idea of a Big-Bang, plus all scientific and philosophical speculations about its nature and what came before will become irrelevant.

This hyperbole actually raises an important question. As discussed in the previous chapters, the techniques we employ to determine the ages of stars depend on the distance and/or the nature of our targets, and the type of observations available, with the common thread that they are all (except for ages derived from radioactive elements) rooted in the results of the theory of stellar evolution. A natural question to address is the following: How accurate are our age determinations? In other words, how close are they to the true values? Obviously, we do not know a priori the true values of these ages, but we can make some educated guesses and set limits to the accuracy of our results, based on the current level of consistency between theoretical stellar models and constraints from asteroseismology (and helioseismology) and spectroscopy (see chapter 5).

The uncertainty on the age of the Sun obtained from radioactive elements in meteorites certainly is very small, on the order of 1%; however, when using radioactive elements for age determinations of other stars (as discussed in chapter 7, this method applies mainly to stars much more metal poor than the Sun) there is a general consensus that the uncertainties are of about 20% at the very least, up to much higher values. This is due to the difficulty in measuring the abundances of the relevant elements from the stellar spectra, and especially the uncertainty on the initial values of the abundance ratios used as chronometers.

All other techniques discussed in the last chapters rely on calculations of stellar evolution models. Here the main theoretical uncertainties are generally related to the term I named f_{mix}, which appears in Eq. 4.5 of chapter 4, one of the equations of stellar structure and evolution. This term includes all processes that change the distribution of chemical abundances inside the stars (convection, gravitational settling, radiative levitation, mixing due to rotation, and their interplay) beside the changes due to nuclear reactions. Additionally, there are recurrent questions about how to accurately calculate the effective temperatures of stellar models with convective envelopes (see the discussion about the Sun in chapter 7).

All these uncertainties affect the lifetimes and evolutionary tracks of models with a given initial mass and chemical composition, hence the position and shape of isochrones in the HRD and CMD, and need then to be folded with the uncertainties of the observations (for example errors in the measured magnitudes, effective temperatures, or chemical abundances of the target stars). In the case of individual stars and star clusters, it is fair to say that, depending on the method, the observational errors and the object under study, uncertainties on their ages are typically in the range between 10% (at best, like for the ages of globular clusters with the highest quality data) and around 30% or slightly more.

When we determine the various epochs of star formation of resolved galaxies we have limits on the time resolution of the SFH. For example, we cannot typically tell whether a generation of stars is 10.5 or 10.9 Gyr old, mainly because photometric errors blur the distinction among isochrones with small differences of age and Z. As mentioned in the previous chapter, the best determinations of the SFH of galaxies have time resolutions of about 1–2 Gyr for ages above 1 Gyr, and 200–400 Myr at younger ages, and metallicity resolutions typically of a factor of two. The uncertainties on stellar evolution models can typically move stars in the recovered SFH by one or at most two time-intervals, and the derived metallicities also by at most one metallicity range.

It is more difficult to estimate the effect of model uncertainties on the derived SFH of unresolved populations. In this case, also the uncertainties on the stellar spectra (see chapter 6) that are added up to calculate the theoretical integrated spectrum of a single-age, single-metallicity population play an important role, because the shapes of a large number of spectral lines must be predicted accurately[1].

Intrinsic uncertainties by a factor up to two on the ages, especially for young (below 1 Gyr) and intermediate-age populations (ages of a few Gyr) are possible, depending on the true SFH of the target galaxy and the quality of the data. This number is indeed comparable to a typical age resolution of 0.2 dex in logarithmic scale (a factor about 1.6) for SFHs derived from good quality integrated spectra of galaxies (see chapter 9). Uncertainties by the same factor on the derived metallicities are also likely.

Our current inability to predict the morphology of the horizontal branch of old populations (see chapters 4 and 7) has also a major effect on some features of the integrated spectra (and on the integrated colours), like for example the H_β absorption feature. We are not able yet to isolate the effect of HB stars in the integrated light[2], hence we have a problem with old and especially metal-poor unresolved populations. If the HB in the target population is bluer than the corresponding

---

[1]The detailed shape of spectral lines does not play such an important role in the calculations of bolometric corrections, because these depend mainly on the average trend of the brightness over tens and hundreds of nanometres, not by the exact shapes of features with sizes on the order of 0.1 nm or less.

[2]There have been attempts to do so, also by myself with Susan Percival a few years ago, which have provided some partial hints at a solution.

branch of the isochrones, we will measure (with spectra and colours) a much younger age than the true one, up to several Gyr. I wrote metal-poor because for old ages and metallicities about a factor of 5 less than solar and higher, the HB is predicted to be always populated close to its red end, for any reasonable uncertainty on the value of the mass lost on the RGB. Given that especially in massive galaxies most of the stars turn out to be in this metallicity range, the problem with the HB morphology affects mainly the ages of unresolved metal-poor globular clusters and, potentially, unresolved dwarf galaxies, which are more metal-poor than massive galaxies.

As mentioned before, refinements to our clocks for maximizing the accuracy of stellar ages require improvements mainly in the description of mixing inside stars. Asteroseismology is helping us and will continue to do so[3] thanks to its power to probe the internal structure of stars. As already mentioned in chapter 4, ideally we could overcome the current limitations of our description of f_{mix} by modelling a star in three dimensions, solving the (complex) equations of hydrodynamics, appropriate to describe the flow of liquids and gases. This way, all mixing processes (and likely also the rates of mass-loss from the surface) come out 'naturally' from the solution of the equations, without the need to model them with a term like f_{mix}. Unfortunately, this is impossible to do today, and will remain unachievable for the foreseeable future. In very simple words, the detailed behaviour of the stellar gas across the whole star is unpredictable in its details even with the most advanced computing tools at our disposal today. However, the ever-increasing power of computational facilities is enabling us to model ever more realistically and in three dimensions small regions within a stellar model. The hope is that the results of these calculations can provide us with guidelines and general simple laws applicable to the whole star, to improve the mathematical description of f_{mix} in our full model calculations.

We also saw in chapter 9 that when studying the integrated spectra of unresolved populations, the set of stellar spectra assigned to each point along an isochrone is a source of uncertainty. In parallel with continuous improvements in the calculations of theoretical stellar spectra, an alternative avenue is to use those of real stars, with known chemical

---

[3]For example, the PLATO (PLAnetary Transits and Oscillations of stars) satellite due to be launched in 2026, will investigate the seismic activity in stars, in addition to its main aim of discovering Earth-like planets in the habitable zone around stars similar to the Sun.

composition, effective temperature, and surface gravity (determined from spectroscopy). These observed spectra, when very high-quality, will obviously reproduce accurately all absorption features; however, they are necessarily restricted to nearby stars – they are biased towards the SFH of the solar neighbourhood and contain mainly old and young stars around solar metallicity – which do not cover all evolutionary phases for all possible ages and metallicities needed to probe the full range of stellar populations in external galaxies [27].

10.2 RECEDING HORIZONS

At the beginning of the seventeenth century, Hans Lipperhey, Zacharias Janssen and Jacob Adriaenszoon, developed independently in the Netherland a novel optical device, which allowed things at a great distance to be seen as if they were nearby.

In 1609 Galileo Galilei became aware of this new device through his friend Paolo Sarpi, designed his own one, and pointed it skyward, starting a revolution in astronomy[4]. The new device got the name we use today on 14 April 1611, at a banquet in honour of Galileo hosted by Prince Frederico Cesi, founder and president of the 'Accademia dei Lincei' a society – still existing – dedicated to scholarly pursuits. Galileo was to be appointed member of this 'Accademia', and the Prince referred to the instrument responsible for Galileo rise to fame as 'teleskopos', meaning 'far-seeing' in Greek, a term coined by one of the other guests, the Greek mathematician Giovanni Demisiani.

These first telescopes were based on lenses to magnify objects (like the glasses I am using to write this book on my laptop computer) and are called 'refracting telescopes' (or 'refractors'): Galileo first telescope had an objective lens with a diameter of about 4 centimetres, and magnified objects by about 20 times. In 1668 Isaac Newton introduced a new design using mirrors, the 'reflecting telescope' (or 'reflector') and over the next centuries both designs evolved to produce ever more powerful instruments. Refractors reached the peak of their power in the nineteenth century with the construction of telescopes with lenses reaching one metre in diameter, but this size proved to be the largest that could be practically constructed. Eventually, the reflector design

---

[4]The first astronomical observations with this new instrument were however made by Thomas Harriot, who observed the moon in the same year.

became the standard, because mirrors of increasingly large diameter could be built.

Together with the construction of ever more powerful telescopes, it has been also crucial the development of means to record images and spectra in a repeatable and consistent way, to detect objects beyond the range of the keenest eyesight, enable more accurate analyses of the data, and consistent intercomparisons among different sets of observations. The mid-nineteenth century saw the development of photography applied to astronomy, with the first photographic image of a star obtained in 1850 by George Bond, and the first photograph of a stellar spectrum showing absorption lines taken in 1872 by Henry Draper. The principle behind photography was the use of light-sensitive chemicals, that once exposed to light react and become opaque to varying degrees depending on the amount of exposure. Photography and photographic plates have been gradually superseded during the twentieth century by detectors that convert the photons from the starlight into an electric current, and allow more precise measurements of brightness. Today they are based on semiconductors – as those we use to take pictures with our mobile phones – and different types of detectors can be used to observe the whole wavelength range from X-rays to the optical and infrared, well beyond the capabilities of photographic plates.

The last century has also witnessed the development of radiotelescopes capable to record signals at radio wavelengths beyond the infrared, and space telescopes that avoid the absorbing (particularly severe for X-rays and large sections of the infrared wavelength range) and blurring effects of the Earth's atmosphere. Finally, the twentieth century also ushered the era of 'adaptive optics' for ground-based telescopes, the ability to partially correct the distortions of astronomical images due to the passage of light through the Earth's atmosphere, by producing tiny deformations in the telescope mirror.

At the time of writing, astronomers have at their disposal 18 ground-based telescopes with diameters in the range between 6 and 10 metres, plus several more instruments with smaller diameters, dozens of radiotelescopes, over 20 space-based telescopes, among which the Hubble Space Telescope (with a 2.4-metre mirror) has been key for the determination of stellar ages in star clusters and Local Group galaxies. In addition, the *Gaia* (Global Astrometric Interferometer for Astrophysics) satellite is currently providing us with photometry and parallaxes (and motions) of over a billion stars in the Milky Way and nearby galaxies. About 50 million stars of the Milky Way have now a parallax accuracy that leads to a 10% (or better) precision in their distances. The accurate distances provided by these parallaxes have led to exquisitely detailed CMDs, and better determinations of ages and SFH of stars in our galaxy.

It is this enormous development of the astronomical instrumentation since the invention of the telescope that has made it possible possible to employ the theory of stellar evolution to determine the ages of stars in the Milky Way and beyond. Just to give a representative number to quantify this progress, a 10-metre telescope like each of the two Keck telescopes at Mauna-Kea in Hawaii, has over 60,000 times the light-gathering power of Galileo telescope, meaning it can detect over 60,000 times more photons per second.

The next generation of observing facilities will improve the precision of current age and SFH determinations by providing better quality data, and will extend the study of stellar ages to more distant targets. In autumn 2021 the James Webb Space Telescope (JWST) should be launched, an upgrade of the Hubble Space Telescope with its 6.5-metre mirror, operating in the infrared. Instead of orbiting the Earth at a distance of 570 Km like the Hubble Space Telescope, JWST will sit 1.5 million Km away in the opposite direction to the Sun, in a position called second Lagrange point. As Earth orbits the Sun, JWST will follow the motion of the planet by remaining in the same spot with relation to both Earth and the Sun. In this way, it can deploy a solar shield to block at the same time the light from Sun, Earth, and Moon, and stay cool, which is very important for its infrared detector. After JWST, the next major new facility relevant to the age of stars to come online around 2025 is the Extremely Large Telescope (ELT), with a diameter of 39 metres (the mirror will actually be made of 798 hexagonal mirror segments), much bigger than the length of a basketball court. It is being built in the Atacama desert in Chile and will be the largest near-infrared telescope in the world, its light-gathering power reaching about 1,000,000 times that of Galileo first telescope, and will provide images 15 times sharper than those from the Hubble Space Telescope.

Both ELT and JWST will work in the infrared, and the CMDs of the resolved stellar populations will appear a bit different in shape from the optical ones we have seen in the previous chapters[5]. Despite this,

---

[5]Compared to optical filters, infrared filters record the brightness of different sections of the stellar blackbody spectra.

we can apply the same techniques to determine ages and SFHs of the target populations.

Observations with both JWST and ELT will provide us with deep CMDs and improved SFHs of galaxies in the Local Group, as well as CMDs down to the oldest turn-off for objects in the outskirts of a few nearby galaxy groups (the M 81 group, NGC 253 group, and Centaurus A group). We will also be able to build CMDs of the brightest stars populating galaxies (including elliptical galaxies, which are lacking in the Local Group) in the Virgo cluster, like bright MS and core He-burning stars of young populations, as well as bright RGB and AGB stars of older populations, to put strong constraints on both their recent and early SFHs. Spectroscopy with the ELT will help this work, by providing measurements of chemical abundances in bright stars in the Local Group and beyond to the Centaurus A group.

With these new facilities, we will also be able to study and determine ages of unresolved stellar populations in massive galaxies to high redshifts, and identify galaxies about 100–300 million years after the Big-Bang, during the first stages of their formation. These future studies of resolved and unresolved galaxies will return a more comprehensive picture of the development of stellar populations in galaxies over time, and provide more stringent tests for the theories of galaxy formation and evolution.

Another main science driver of these new instruments is the study of planetary systems during their formation from discs around nearby very young stars[6], and the study of planetary atmospheres of Earth-like planets in the habitable zone, to search for the presence of molecules indicative of biological processes. Likely in a not-too-distant future, maybe thanks to JWST and ELT, we'll finally discover signatures of biological activity in a sample of exoplanets, and determinations of the ages of their host stars will be an absolutely crucial piece of information to be able to study the timeframe for the development of life across our galaxy.

---

[6]It should now be clear why JWST and ELT will work in the infrared. The light of the bright, hot, and young stars in high-redshift stellar populations, formed when the universe was young, was emitted mainly at ultraviolet and optical wavelengths but it has been stretched by the expansion of the universe, and we see it today as infrared light. Another reason is that planets and stars form in clouds of gas and dust, that cause very high extinction at optical and shorter wavelengths. Infrared light crosses these clouds much more easily because extinction is lower at longer wavelengths, and allows us to see inside them.

While I was considering these developments, I couldn't help but thinking of what Agnes Clerke wrote at the end of the nineteenth century about the new discipline of *astronomical physics*, concerned about the nature of the *heavenly bodies*, not just their movements: *It is not too much to say that a new birth of knowledge has ensued. The astronomy .... placed by Comte at the head of the hierarchy of the physical science was the science of the movements of the heavenly bodies. And there were those who began to regard it as a science which, from its very perfection, had ceased to be interesting – whose tale of discoveries was told, and whose farther advance must be in the line of minute technical improvements, not of novel and stirring disclosures. But the science of the nature of the heavenly bodies is one only in the beginning of its career. It is full of the audacities, the inconsistencies, the imperfections, the possibilities of youth. It promises everything; it has already performed much; it will doubtless perform much more.*

Very prescient words indeed!

Assorted (astro-)Physics

I have always found fascinating the additional sections that often follow the last chapter of scientific and historical essays. They usually include biographies of the main characters in the book, chronologies, maps, extended notes to the chapters, summaries of basic scientific concepts. When I was a teenager, I often liked to have a peek at those pages before the actual book.

Here I have gathered a summary of basic concepts and background to the various chapters. I have preferred to group these notes into an appendix, that can be readily consulted while reading the book.

You'll find summarized here the basics of the Standard Model of particle physics, the thermodynamics relevant to understand how stars work, short discussions about parallax distances, spectroscopic parallaxes (which actually have nothing to do with the parallax), and how we determine mass and radius (and distance) of stars in eclipsing binary systems.

A.1 CHEMICAL ELEMENTS, ELEMENTARY PARTICLES AND THE FOUR FUNDAMENTAL FORCES OF NATURE

Probably the most basic question we can ask ourselves is: What are we made of? Around the fifth century BC the Greek philosophers Leucippus of Miletus and Democritus of Abdera developed the idea that all matter is made up of tiny, indivisible particles called atoms, a term coming from the Greek word *atomos*, meaning *indivisible*. These atoms were considered to be too small to be seen, unchanging, indestructible, with no internal structure, coming in an infinite variety of shapes and sizes, which accounted for the different kinds of matter. After being disputed by the very influential Aristotle, whose teachings shaped the views of Roman Catholic theologians, the concept of the atom was revived in the nineteenth century. The chemist, physicist, and meteorologist John Dalton first suggested (probably influenced by the views of the chemist Bryan Higgins) that all chemical elements in nature are made of tiny particles called atoms, which cannot be partitioned, created, or destroyed. All atoms of a given element are identical in size, mass, and other properties, characteristic of each specific element, and in chemical reactions atoms are combined, separated, or rearranged. Dalton also published a table of relative atomic weights for six elements – hydrogen, oxygen, nitrogen, carbon, sulphur and phosphorus – relative to the weight of an atom of hydrogen, conventionally taken as 1.

Our modern picture of atoms started to coalesce in 1897, when Joseph Thomson discovered the electron, followed by the discovery of the proton in 1911 thanks to Ernest Rutherford, and the neutron by James Chadwick in 1932. Atoms are made of a nucleus that contains protons (particles with a positive electric charge) and neutrons (electrically neutral, with almost the same mass as the protons). Around the nucleus move the electrons, which carry a negative electric charge and are about 2000 times less massive than protons and neutrons. The number of electrons in an atom is equal to the number of protons, to ensure electrical charge neutrality.

Today we know that typically 92 chemical elements occur in nature. Each chemical element is made of atoms with a given value of the 'atomic number', defined as the number of protons in the nucleus; elements with atomic number up to 92, corresponding to uranium, can be found in nature, together with minute amounts of elements up to atomic number 98. Higher atomic numbers correspond to man-made elements. Pretty much the same elements found on Earth are also observed in the spectra of stars and galaxies, although the ratios of their abundances are not the same everywhere in the universe.

A chemical element is identified by its unique atomic number, but there can be different 'versions' of the same element, or 'isotopes', which differ by the number of neutrons in the nucleus. For example, hydrogen (H) nuclei are usually made of one proton, with no neutrons. But there exist two other isotopes of hydrogen, named deuterium (D or ^{2}H) and tritium (^{3}H – where the number attached to the hydrogen symbol denotes the number of particles in the nucleus, the sum of protons and neutrons) with one and two neutrons in the nucleus, respectively. In general, apart from hydrogen, the number of protons and neutrons in the nucleus is typically the same (for example the most abundant isotope of carbon has 6 protons and 6 neutrons, hence it is denoted as ^{12}C) and isotopes with extra neutrons are generally radioactively unstable[1], meaning that the neutrons gradually change into protons by emitting an electron in the so-called β decay process (see below). As a result, an unstable isotope of a certain element transforms into a different element, because the number of protons in the nucleus has changed.

In addition to the atomic number, chemical elements are assigned also an 'atomic weight', defined as the ratio of the average mass of its atoms – the average account for the percentages of all its isotopes occurring in nature – to a standard. This standard is defined as one-twelfth of the mass of an atom of the carbon isotope ^{12}C. For example, the atomic weight of helium is 4.0026, because 99.9998% of helium in nature is the isotope ^{4}He (two protons and two neutrons in the nucleus) and 0.0002% is the isotope ^{3}He (only one neutron in the nucleus). The atomic weight of carbon itself is 12.01, that accounts for the small percentage of ^{13}C (about 1%) that occurs in nature.

This shows that atoms do have a structure, they are not the building blocks of matter, but actually, even protons and neutrons are made of more fundamental subunits, as briefly discussed in what follows.

The extraordinary amount of experimental and theoretical work that thousands of physicists have carried out since the 1930s has co-alesced in the early 1970s into our current picture of the most funda-mental structure of matter, the so-called Standard Model of particle physics [65]. Let's be clear that here 'matter' means everything in the universe but dark matter and dark energy, whose presence has been inferred from astrophysical observations. Given that dark matter and dark energy make about 95% of the universe, and that gravity (as we will see soon) needs to be treated separately, it is perhaps an under-statement to state that the Standard Model is incomplete. However, from the point of view of particle physics experiments, it is fair to say that the Standard Model performs remarkably well. According to this model, the matter is made of a few (although not very few) building

---

[1]This is not always true, as the case of the stable carbon isotope ^{13}C with 6 protons and 7 neutrons in the nucleus.

blocks called fundamental particles, which follow laws governed by four fundamental forces.

These fundamental particles occur in two basic types, 'quarks' and 'leptons'. Each of these two groups comprises six particles, or 'flavours' (not a term you easily imagine being associated with theoretical physics). Quarks are named *up, down, charm, strange, top,* and *bottom*, whilst leptons are the electron, electron neutrino, muon, muon neutrino, tau, and tau neutrino. All quarks and all leptons but the neutrinos have an electric charge, and quarks are characterized also by another property, the 'colour charge' (they can come in three different colour charges, named green, blue, and red), that leptons do not have. Both quarks and leptons are 'fermions', a class of particles whose common property is essentially that no two identical objects may share the same energy.

Quarks are bound together into groups of three (triplets) and two (doublets) to make other particles, the 'baryons' and 'mesons', respectively. In the case of an atom's nucleus, its constituents are baryons; protons are made of two *up* quarks and one *down* quark, while neutrons are composed of one *up* quark and two *down* quarks.

In addition, every quark and lepton has its so-called antiparticle. An antiparticle has the same mass as the corresponding particle, but the opposite physical charge, like electric charge[2] or colour charge[3]. When a particle and its antiparticle interact, they transform into photons (a process called annihilation) whose energy is equal to $2mc^2$, where m is the mass of the particle and c the speed of light (see below for more details). Conversely, whenever photons have an energy equal to $2mc^2$, where m is the mass of a generic particle, this latter can be created together with its antiparticle at the expenses of the photon energy. This particle-antiparticle pair eventually annihilate, giving back photons.

These fundamental particles, and the more complex particles they are assembled into, interact with each other according to the four fundamental forces that are at play in the universe: the *strong* force, *weak* force, *electromagnetic* force, and *gravity*.

Gravity is the weakest force, but it has an infinite range, like the electromagnetic force that is however many times stronger than gravity.

---

[2] As an example, the antiparticle of the electron is the positron, that we can think of as an electron with a positive electric charge.

[3] The opposite of the green, blue and red colour charges are named, perhaps not surprisingly, anti-green, anti-blue, anti-red. Regarding the difference between neutrinos and antineutrinos the situation is a bit more complicated.

The strong force is effective only over a distance of about 10^{-15} m (roughly the size of a proton) and the weak force only up to about 10^{-18} m. Over the short distances where the four fundamental forces are all effective, the strong force is about a factor 100 stronger than the electromagnetic force, about 100000 times stronger than the weak force, and about 42 orders of magnitude (equal to the number one followed by 42 zeroes) stronger than the gravitational force.

If we leave aside gravity for the moment, the electromagnetic force acts among particles that carry an electric charge (all quarks and leptons except for the neutrinos), the strong force involves particles with a colour charge (only quarks and their bound systems, baryons, and mesons), whilst the weak force affects all quarks and leptons. The electromagnetic force can be both repulsive (if acting between electric charges of the same sign) or attractive (between charges of a different sign), holds atoms together (keeps negatively charged electrons tied to the positively charged nuclei), and is responsible for the chemical bonds between atoms which create molecules. It is the strong force that binds the protons and neutrons of the atomic nucleus together, despite the intense repulsion amongst the positively charged protons. The strong force is attractive at distances typical of the atomic nucleus, decreases fast at larger distances, but becomes repulsive on scales smaller than the nucleus. The weak force instead transforms quarks and leptons from one flavour to another, and is responsible for radioactive decays, like the spontaneous decay of free neutrons (free neutrons are not stable and decay after about 15 minutes) into protons, electrons and electron antineutrinos, or the decay of radioactive elements (a neutron in their unstable nuclei transforms into a proton and an electron plus an antineutrino).

How do particles 'feel' these forces? In our current view (extrapolated from the complex mathematics that describes these processes) these forces result from the exchange of carriers, the 'gauge bosons', which transmit the force by travelling between the affected particles. In this way, matter particles transfer discrete amounts of energy by exchanging gauge bosons with each other. An important property of bosons is that, contrary to fermions, two or more identical bosons can share the same energy. The electromagnetic force is carried by the photon, the gluon (there are eight types of gluons) mediates the strong force, and the W (positive and negative charged) and Z (electrically neutral) particles mediate the weak force. Gluons and photons have zero mass (and zero electrical charge) and travel in vacuum at the speed of light.

In a way, the interaction between two particles is like when two friends meet on the street. They exchange words (usually) carrying information (the words, in this case, take the role of gauge bosons) and their status is altered (for example, each friend learn something new) by this encounter (interaction) because of the new information conveyed through their words.

Gauge bosons exchanged during an interaction should be thought of as 'virtual', in the sense that they exist only for a fleeting moment, qualitatively similar to the words exchanged when people meet. Energy is borrowed (see below for the concept of matter-energy equivalence) for a short time to transfer the force via gauge bosons and then it is paid back to the matter. For example, when the neutron decays one *down* quark transforms to an *up* quark which has a lower mass. In the process, a virtual W^- gauge boson is created that then decays into an electron and its antineutrino.

In addition to six quarks, six leptons and five types of gauge bosons (photon, gluon, positive and negative W, Z), there is one more fundamental particle, predicted theoretically in 1964 and discovered as recently as 2012, the so-called Higgs boson. All quarks, leptons, the W and Z particles acquire their masses by interacting with Higgs bosons that permeate space. The variety of masses of the fundamental particles arise because different particles have different strengths of interaction with the Higgs particles. In simple terms, the interaction with Higgs bosons makes harder for the particle to travel, and since mass can be seen as resistance to motion (we experiment this in everyday life when we push heavy objects), in a sense they acquire a non-zero mass from this process. Going back to the picture of interaction sketched before, if I am a particle (for example a quark *up*) walking on my way to work, the larger the number of times I have to speak (interaction) to people around me (the Higgs bosons permeating space), the slower will be my progress towards my workplace (the higher my mass, that is resistance to motion).

This very concise, qualitative, and incomplete description of the fundamental particles and forces summarizes the Standard Model of particle physics. It successfully merges both the laws of quantum mechanics and the special theory of relativity, explaining almost all experimental results in particle physics so far, and has precisely predicted a wide variety of observed phenomena.

The fourth fundamental force that has been left out of this discussion is gravity, probably the most familiar force in our lives that acts among all particles. Despite decades-long attempts to describe gravity in the same way as the other three forces, no one has yet managed to provide a description of gravity that is mathematically compatible with the Standard Model. Einstein general theory of relativity [35] is extremely successful at describing gravity, but it is profoundly different from the mathematical apparatus of the Standard Model. General relativity envisages gravity as the effect of the curvature of space around mass (and energy). The motion of particles follows the curvature of space – like our train travels on this planet follow the spherical curvature of Earth – hence they are attracted by the gravitational field of a massive object, because mass curves the surrounding space, like a heavy sphere on an elastic sheet. No force carrier mediates the gravitational interaction: Gravity is a geometrical effect, although the presence of a 'graviton' is hypothesized by theories attempting to describe gravity in the same way as the other fundamental forces.

From a pragmatic point of view, the incompatibility of gravity with the other three forces is usually not an issue. At the small distances of particle interactions, the effect of gravity is negligible compared to the other forces. On large scales matter is usually electrically neutral, gravity dominates, and can be treated independently of the other forces. Problems, however, arise in the description of the very early universe and the centres of black holes, where the distances involved require the combined treatment of gravity and the other interactions.

A.1.1 Mass-Energy Equivalence

As mentioned before, the Standard Model complies with the laws of both quantum mechanics and special relativity. In particular, the mass-energy equivalence, one of the main results of special relativity, is automatically accounted for in the mathematical formalism of the model. Let's consider briefly the implications of this equivalence, given by the famous formula $E = mc^2$, where on the left-hand side we have an energy (a measure of the capacity of an object to do work, that is to produce movement) and on the right-hand side the product of a mass times the square of the speed of light c. This relation means that if we consider a particle of mass m, a huge amount of energy E is associated just to its mass. The amount of energy equivalent to a mass of 1 gram is slightly more than the energy released by the devastating atomic bomb that exploded over Hiroshima in 1945.

The first important implication of this equivalence involves gravity. The force of gravity acts between masses, but if mass is equivalent to energy, also energy can cause gravity, hence curvature of space[4]. The second implication, crucial in stellar physics, is that mass can be transformed into energy. In nuclear reactions, like the conversion of hydrogen to helium in the Sun, the mass of the reaction products is always slightly lower than the sum of the masses of the interacting nuclei. The amount of mass 'lost' has become energy according to $E = mc^2$. The third implication is even more surprising: Energy can be used to create particles. Let's imagine to have very energetic photons (with very short wavelength), like in the first moments after the Big-Bang. As an example, if the energy of these photons is equal to twice the mass of the electron multiplied by the square of c, electron-positron pairs can be created, which will eventually annihilate and transform back into photons.

A.2 THERMODYNAMICS

Thermodynamics is the branch of physics that deals with the transfer of energy from one place to another and from one form to another, and it is crucial to understand the evolution of physical and biological systems, including of course stars. In a slightly more formal way, we can say that thermodynamics studies the exchange of energy between a system and the environment, to identify the conditions for the spontaneous evolution of the system itself.

These definitions contain the key terms 'energy' and 'system', or also 'thermodynamic system'. As already said, energy is a measure of the capacity of an object to do work. A common form of energy is the kinetic energy of a moving body (defined as half the value of the object mass multiplied by the square of its velocity), but it is also important the concept of potential energy, that is the energy stored in an object due to its position or state. We can think of it as an energy that has the potential to do work, hence of generating movement. When the position (or the state) of the object changes, the stored energy will be released, and the object will move. A classical example is the gravita-

---

[4]This equivalence is also the reason why photons, although massless, 'feel' the force of gravity.

tional potential energy of a book on a shelf. If the shelf is removed, the book will move and fall to the floor. Another example is the elastic potential energy stored in elastic materials when they are stretched or compressed. A stretched rubber sling has an amount of elastic potential energy that produces movement when the sling is released. Also, a positive electric charge near another positive charge has an electrostatic potential energy, for if left free to move, it would be repelled away from the like charge.

The other keyword is 'thermodynamic system'. A thermodynamic system denotes a portion of space separated from the surrounding environment through a surface (that can be a physical surface or just imaginary); within this enclosed space active processes cause internal transformations and exchanges of matter and/or energy with the environment.

As an example, petrol inside a cylinder of a car engine is a thermodynamic system; the inner surfaces of the cylinder and piston are the surfaces that enclose the system, while the rest of the engine is the environment. A human being can also be considered a thermodynamic system, and our skin is the surface that separates us from the external environment, Here I will consider a much simpler system, useful in the context of stellar physics, a gas in a container.

The condition ('state') of a system at any given time is its 'thermodynamic state', and for our gas in a container the state is identified by the gas chemical composition, temperature (denoted with T), pressure (P), and density (ρ). While chemical composition and density (total mass of the gas divided by its volume, equal to the volume of the container in this example) are basic concepts, the definitions of temperature and pressure are less trivial.

The simplest way to introduce the concept of temperature is to consider a perfect gas (see chapter 3) composed of a large number of particles (atoms or molecules) whose size is negligible compared to their relative distances; they move randomly, often colliding with each other and with the surfaces of the container. Due to their continuous motions the gas particles have a momentum (the product of a particle's mass times its velocity) and kinetic energy. The temperature of a gas is a measure of the average kinetic energy of the particles, due to their random motions. In a hot gas, the average kinetic energy is higher than in a cold gas, and given that the mass of the particles does not change with velocity (unless they are approaching the speed of light), higher temperature means higher particle velocities[5]. Also, the gas particles collide with the interior of the container, producing a force perpendicular to these surfaces. The sum of the forces of all the particles striking the inner surfaces of the container divided by their area is defined to be the pressure. For a given chemical composition, the values of P, T, and ρ of the gas are related by the so-called equation of state. If we fix two of these variables, the equation of state will give us the value of the remaining one. For a perfect gas with a given chemical composition the pressure increases with increasing T when the density is fixed, and also increases with increasing ρ when T is fixed.

Another important quantity that characterizes a thermodynamic system is its 'internal energy' (usually denoted with U). Two terms enter the definition of U. The first one is the sum of the kinetic energy due to the random motions of the particles plus, in case of molecules, the kinetic energy associated to the rotation and vibration of the constituent atoms. For example, in the case of molecules composed of pairs of atoms, the two components rotate about each other and alter their relative distance periodically (vibrations). The second term is the potential energy of the particles (if any), like the potential energy associated to the relative positions of the atoms that make up the molecules, kept together by the electromagnetic interaction.

Whilst the internal energy is essentially the total energy content of a system, 'heat' can be seen as energy in transit, being transferred from a system to another, or between different parts of the same system. When there is no external intervention, the 'second law of thermodynamics' says that heat is always transferred from higher to lower temperature systems (or parts within a system), not in the reverse direction.

Let's consider two gases at different temperatures. When they are mixed together the constituents of the hotter gas lose kinetic energy, and the colder gas gain kinetic energy due to interparticle collisions, until the average kinetic energies of the groups of particles originally in the two separate systems attain the same value. The temperature has become the same for both gases, lower than the one of the originally hotter gas, but higher than the temperature of the cooler gas. During the process, energy (heat) has been transferred from the hotter system to the cooler one, to reach the intermediate equilibrium temperature of

---

[5]This definition of temperature can explain qualitatively also the case of a gas of photons behaving like a blackbody. The energy distribution of the photons is shifted towards higher energies (lower wavelengths) for higher blackbody temperatures. A more general definition of temperature can be found in physics textbooks.

the combined gases. The amount of heat that needs to be transferred to cause a unit temperature change is called 'heat capacity'.

Another fundamental law of thermodynamics relevant to the evolution of stars is the 'first law of thermodynamics': It states essentially that energy can neither be created nor destroyed, just transferred or changed from one form to another. As a result, a system can exchange energy with its surroundings by the transmission of heat and performing work[6]: The net energy exchanged is equal to the change in the internal energy of the system. If the system does not interact with the environment ('isolated' system), its internal energy stays constant.

A.3 PARALLAX DISTANCES

The direct method to determine the distance to a star, and in general to astronomical bodies, is based on 'parallax' measurements, and relies only on basic geometry. The concept of parallax is easy to experiment in our own room. Let's place ourselves at a certain distance from a wall on which, for example, a clock is hung. We close the right eye, and then extend the left arm and pull up the thumb to cover the image of the watch. Now let's close the left eye and open the right one. What we will see with the right eye is that the thumb no longer covers the image of the clock, but it is shifted to the left side of the clock. By measuring the change in the position of the thumb relative to the background static clock, and knowing the distance between our eyes, we can calculate the distance from our face to the thumb using geometry. The longer the arm, the further away the thumb from the face, and the smaller its shift with respect to the clock when we watch with the left eye closed.

The same principle applies to the determination of astronomical distances, as sketched in Fig. A.1. The figure shows the apparent shift in a star position (the thumb in the previous example) with respect to a background of much more distant objects (the clock on the wall), caused by the motion of the Earth around the Sun (opening and closing one eye at a time). If the shift is measured in terms of the angular separation between the positions of the star measured six months apart, half of the separation (denoted with θ in the figure) is called 'parallax'.

To give some reference values of angular separations, let's hold a hand at arm's length and close one eye. If I now make a fist with the

---

[6]For example, a perfect gas confined by a frictionless piston performs work by pushing the piston upwards.

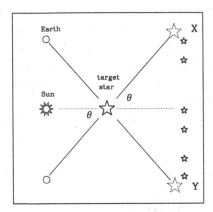

Figure A.1 As Earth orbits the Sun, when we measure the position of a nearby star against a background of distant ones, its location shifts from X to Y when observed six months apart. The angle θ (half the angular separation between positions X and Y) is called parallax. Parallaxes are very small, the largest value is only about 0.77 arcsec.

back of my hand facing me, the width of the fist corresponds approximately to 10 degrees, while the three middle fingers will span about 5 degrees. The average angular diameter of the full Moon is just about 31 arcminutes (arcmin), with 1 arcmin being equal to 1/60 of a degree. Parallaxes of stars are much smaller than these values: The largest parallax is measured for the closest star, Proxima Centauri, and it is only about 0.77 arcseconds (arcsec), whereby 1 arcsec is equal to 1/60 arcmin. With such small angles, a star distance D is given to great accuracy by D=a/θ, where a is the Earth-Sun distance and θ the measured parallax of the target. This formula becomes simply D=$1/\theta$ if the parallax is given in arcseconds and D in parsecs, where 1 parsec is equal to 3.26 light-years and by definition corresponds to a parallax of 1 arcsec.

The small values of the parallax angles explain why parallax distances are generally limited to within the Milky Way. The first stellar distances using parallax measurements were obtained in 1838 by Friedrich Bessel, who measured the parallax of the star 61 Cygni as 0.31 arcsec, whilst almost at the same time Friedrich von Struve and Thomas Henderson measured the parallaxes of Vega and α Centauri, respectively. In recent times, the European Space Agency (ESA) satellite *Hipparcos* has measured between 1989 and 1993 the parallaxes of 120000 stars with an accuracy of 0.001 arcsec, and about 2.5 million stars with lesser accuracy. The ongoing ESA *Gaia* mission is measuring parallaxes of about 1.5 billion stars with an accuracy up to several orders of magnitude better than *Hipparcos* measurements.

A.4 SPECTROSCOPIC PARALLAX

As mentioned in chapter 3, Walter Adams and Arnold Kohlschütter found in 1914 that some spectral lines change their appearance between luminous and faint stars of known distance. Further works found that this is a consequence of the fact that at the same spectral type, meaning at approximately the same surface temperature, denser stellar atmospheres broaden the absorption lines. This has led to an additional classification within a spectral type, according to the 'luminosity class'. The scheme used today was developed and published by William Morgan and Philip Keenan in 1941. In addition to a spectral type like G2 in case of the Sun, stars are assigned an additional Roman numeral: I (stars in this class are named supergiants, often divided into the further classes Ia-0, Ia, and Ib in descending order of brightness), II (bright giants), III (giants), IV (subgiants), V (dwarfs).

This sequence traces the increase of atmospheric densities, related to the surface gravity, hence follows a progression towards decreasing radii. Higher surface gravities correspond to more compact, smaller radius and fainter (at a given surface temperature) stars. With this classification the Sun is a G2V dwarf, Betelgeuse a M2 Ib supergiant, Sirius a A1V dwarf, just to give some examples. Figure A.2 displays a CMD of stars in the solar neighbourhood (stars with distances up to about 100 pc) including the corresponding spectral types and luminosity classes, and shows how these luminosity classes map onto the position of stars in the CMD. Similar mapping holds in a Russell-type HRD with spectral type on the horizontal axis and absolute magnitude on the vertical one. This means that the spectral type (or colour) and luminosity class measured from the spectrum of a generic star of unknown distance can determine its rough location in the diagram and, in turn, provide a rough estimate of its absolute magnitude, hence its distance modulus. This approximate method to determine a star's distance modulus is called spectroscopic parallax, even though it doesn't involve any parallax measurements. It should actually be called

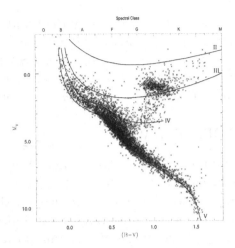

Figure A.2 CMD of stars in the solar neighbourhood with absolute magnitudes derived from parallax distances [123]. Spectral types and luminosity classes (Roman numerals) are also marked. Luminosity class I stars – brighter than the brightest stars in the figure – are missing from this diagram (courtesy of David Hyder).

spectroscopic distance, but the term spectroscopic parallax is still used today.

A.5 MASS, RADIUS, AND DISTANCE OF STARS IN ECLIPSING BINARY SYSTEMS

An eclipsing binary star is a system of two stars orbiting around their centre of mass, whose orbital plane lies nearly in the line of sight of the observer[7]. The approximate alignment of the orbital plane of an eclipsing binary with the line of sight means that, when the binary is seen from Earth, the component stars pass in front of each other periodically causing eclipses, hence the name eclipsing binaries.

The two components are generally too close to be detected as separate objects, and therefore we can only measure the total brightness due to the added luminosity of both stars. If we study the variation of

---

[7]To visualize the concept of centre of mass in the case of a binary star, we can think of two kids playing on a see-saw. If we want to have the load balanced on both sides of the see-saw, we need to carefully choose the position of the pivot point, which has to be closer to the heavier child. When equilibrium is achieved, the position of the pivot point along the bar is the centre of mass of the system.

brightness as a function of time (the 'light curve' of the system), we will see that periods of constant brightness are interrupted by drops which repeat at regular intervals. These drops in luminosity correspond to the transit of one star in front of the other, which shields part of (or all) the light of the eclipsed component. This typically happens twice during the orbit, one for each component eclipsing the other one; the size of the brightness drops depend on the relative brightness of the two stars, and the fraction of the surface of the occulted star that is hidden.

From the light curve, we can determine the orbital period of the system, by simply measuring the time between two repeated sets of eclipses. If we take spectra of the eclipsing binary during one orbital period, we will see spectral lines shifting back and forth because of the Doppler effect due to the motion of each of the two components either towards us, or away from us. In this way we can measure the orbital velocities of the two stars which, coupled to the knowledge of the period, allow us to determine their masses using simple laws from mechanics. By measuring the duration of the brightness dips in the light curve we can also measure how long each eclipse lasts; together with the knowledge of the orbital velocities of the two components, this allows us to determine their radii (hence the surface gravities from the knowledge of masses and radii).

Once the individual radii are determined, we can disentangle the spectra of the two components in our measurements, and derive their individual T_{eff} by comparison with theoretical spectra. When the effective temperature and radius of the components are known, we can then determine their luminosities. The absolute magnitude of each component star in the photometric filter of the observed light curve can be calculated from the luminosity and appropriate bolometric corrections[8].

After these individual absolute magnitudes are added[9], a comparison with the apparent magnitude of the system out-of-eclipse gives us the distance, provided the extinction along the line of sight is known.

---

[8]To compute the bolometric corrections we need to have an estimate of the metallicity of the binary (both stars should have the same initial chemical composition because they formed from the same parent molecular cloud), in addition to the knowledge of T_{eff} and surface gravity.

[9]The combined absolute magnitude $M_{X,bin}$ of the binary in a generic filter X is equal to $M_{X,bin} = -2.5 \log(10^{-0.4 M_{X,1}} + 10^{-0.4 M_{X,2}})$, where 1 and 2 are the two components.

Bibliography

[1] Jairo A. Alzate, Gustavo Bruzual, and Daniel J. Díaz-González. Star Formation History of the Solar Neighbourhood as Told by Gaia. *Monthly Notices of the Royal Astronomical Society*, 501(1):302–328, January 2021.

[2] A. Aparicio, C. Gallart, C. Chiosi, and G. Bertelli. Model Color-Magnitude Diagrams for Hubble Space Telescope Observations of Local Group Dwarf Galaxies. *Astrophysical Journal Letters*, 469:L97, October 1996.

[3] H. C. Arp, W. A. Baum, and A. R. Sandage. The HR Diagrams for the Globular Clusters M 92 and M 3. *Astronomical Journal*, 57:4–5, April 1952.

[4] L. H. Auer and N. J. Woolf. Mass Loss and the Formation of White-Dwarf Stars. *Astrophysical Journal*, 142:182, July 1965.

[5] John N. Bahcall, M. H. Pinsonneault, and G. J. Wasserburg. Solar Models with Helium and Heavy-Element Diffusion. *Reviews of Modern Physics*, 67(4):781–808, October 1995.

[6] Dana S. Balser. The Chemical Evolution of Helium. *Astronomical Journal*, 132(6):2326–2332, December 2006.

[7] Nate Bastian and Carmela Lardo. Multiple Stellar Populations in Globular Clusters. *Annual Review of Astronomy and Astrophysics*, 56:83–136, September 2018.

[8] Timothy R. Bedding, Benoit Mosser, Daniel Huber, Josefina Montalbán, Paul Beck, Jørgen Christensen-Dalsgaard, Yvonne P. Elsworth, Rafael A. García, Andrea Miglio, Dennis Stello, Timothy R. White, Joris De Ridder, Saskia Hekker, Conny Aerts, Caroline Barban, Kevin Belkacem, Anne-Marie Broomhall, Timothy M. Brown, Derek L. Buzasi, Fabien Carrier, William J. Chaplin, Maria Pia di Mauro, Marc-Antoine Dupret, Søren Frandsen,

Ronald L. Gilliland , Marie-Jo Goupil, Jon M. Jenkins, Thomas Kallinger, Steven Kawaler, Hans Kjeldsen, Savita Mathur, Arlette Noels, Victor Silva Aguirre, and Paolo Ventura. Gravity Modes as a Way to Distinguish Between Hydrogen- and Helium-Burning Red Giant Stars. *Nature*, 471(7340):608–611, March 2011.

[9] L. R. Bedin, M. Salaris, G. Piotto, S. Cassisi, A. P. Milone, J. Anderson, and I. R. King. The Puzzling White Dwarf Cooling Sequence in NGC 6791: A Simple Solution. *Astrophysical Journal Letters*, 679(1):L29, May 2008.

[10] Luigi R. Bedin, Ivan R. King, Jay Anderson, Giampaolo Piotto, Maurizio Salaris, Santi Cassisi, and Aldo Serenelli. Reaching the End of the White Dwarf Cooling Sequence in NGC 6791. *Astrophysical Journal*, 678(2):1279–1291, May 2008.

[11] Luigi R. Bedin, Maurizio Salaris, Giampaolo Piotto, Ivan R. King, Jay Anderson, Santi Cassisi, and Yazan Momany. The White Dwarf Cooling Sequence in NGC 6791. *Astrophysical Journal Letters*, 624(1):L45–L48, May 2005.

[12] Eric F. Bell and Roelof S. de Jong. The Stellar Populations of Spiral Galaxies. *Monthly Notices of the Royal Astronomical Society*, 312(3):497–520, March 2000.

[13] Travis A. Berger, Daniel Huber, Eric Gaidos, Jennifer L. van Saders, and Lauren M. Weiss. The Gaia-Kepler Stellar Properties Catalog. II. Planet Radius Demographics as a Function of Stellar Mass and Age. *Astronomical Journal*, 160(3):108, September 2020.

[14] Gianpaolo Bertelli, Mario Mateo, Cesare Chiosi, and Alessandro Bressan. The Star Formation History of the Large Magellanic Cloud. *Astrophysical Journal*, 388:400, April 1992.

[15] M. Bolte and C. J. Hogan. Conflict over the Age of the Universe. *Nature*, 376:399–402, August 1995.

[16] E. Brocato, V. Castellani, and A. Piersimoni. The Age of the Globular Cluster M68. *Astrophysical Journal*, 491(2):789–795, December 1997.

[17] E. Margaret Burbidge, G. R. Burbidge, William A. Fowler, and F. Hoyle. Synthesis of the Elements in Stars. *Reviews of Modern Physics*, 29(4):547–650, January 1957.

[18] B. W. Carney. The Absolute Magnitudes of Field RR Lyrae Stars and Globular Cluster Ages. *Memorie della SocietaÁstronomica Italiana*, 63:409–430, 1992.

[19] Santi Cassisi and Maurizio Salaris. Multiple Populations in Massive Star Clusters under the Magnifying Glass of Photometry: Theory and Tools. *Astronomy and Astrophysics Review*, 28(1):5, July 2020.

[20] Joseph W. Chamberlain and Lawrence H. Aller. The Atmospheres of A-Type Subdwarfs and 95 Leonis. *Astrophysical Journal*, 114:52, July 1951.

[21] W. J. Chaplin, S. Basu, D. Huber, A. Serenelli, L. Casagrande, V. Silva Aguirre, W. H. Ball, O. L. Creevey, L. Gizon, R. Handberg, C. Karoff, R. Lutz, J. P. Marques, A. Miglio, D. Stello, M. D. Suran, D. Pricopi, T. S. Metcalfe, M. J. P. F. G. Monteiro, J. Molenda-Żakowicz, T. Appourchaux, J. Christensen-Dalsgaard, Y. Elsworth, R. A. García, G. Houdek, H. Kjeldsen, A. Bonanno, T. L. Campante, E. Corsaro, P. Gaulme, S. Hekker, S. Mathur, B. Mosser, C. Régulo, and D. Salabert. Asteroseismic Fundamental Properties of Solar-type Stars Observed by the NASA Kepler Mission. *Astrophysical Journal Supplement*, 210(1):1, January 2014.

[22] Benjamin Charnay, Eric T. Wolf, Bernard Marty, and François Forget. Is the Faint Young Sun Problem for Earth Solved? *Space Science Reviews*, 216(5):90, July 2020.

[23] Jieun Choi, Aaron Dotter, Charlie Conroy, Matteo Cantiello, Bill Paxton, and Benjamin D. Johnson. Mesa Isochrones and Stellar Tracks (MIST). I. Solar-scaled Models. *Astrophysical Journal*, 823(2):102, June 2016.

[24] J. Christensen-Dalsgaard and D. O. Gough. Towards a heliological inverse problem. *Nature*, 259(5539):89–92, January 1976.

[25] Joergen Christensen-Dalsgaard. Solar Structure and Evolution. *arXiv e-prints*, page arXiv:2007.06488, July 2020.

[26] Roberto Cid Fernandes, Abílio Mateus, Laerte Sodré, Grażyna Stasińska, and Jean M. Gomes. Semi-Empirical Analysis of Sloan Digital Sky Survey Galaxies - I. Spectral Synthesis Method. *Monthly Notices of the Royal Astronomical Society*, 358(2):363–378, April 2005.

[27] Paula R. T. Coelho, Gustavo Bruzual, and Stéphane Charlot. To Use or not to Use Synthetic Stellar Spectra in Population Synthesis Models? *Monthly Notices of the Royal Astronomical Society*, 491(2):2025–2042, January 2020.

[28] J. Crampin and F. Hoyle. On the Change with Time of the Integrated Colour and Luminosity of an M67-Type Star Group. *Monthly Notices of the Royal Astronomical Society*, 122:27, January 1961.

[29] T. J. L. de Boer, E. Tolstoy, V. Hill, A. Saha, K. Olsen, E. Starkenburg, B. Lemasle, M. J. Irwin, and G. Battaglia. The Star Formation and Chemical Evolution History of the Sculptor Dwarf Spheroidal Galaxyâ. *Astronomy & Astrophysics*, 539:A103, March 2012.

[30] T. J. L. de Boer, E. Tolstoy, V. Hill, A. Saha, E. W. Olszewski, M. Mateo, E. Starkenburg, G. Battaglia, and M. G. Walker. The Star Formation and Chemical Evolution History of the Fornax Dwarf Spheroidal Galaxy. *Astronomy & Astrophysics*, 544:A73, August 2012.

[31] Armin J. Deutsch. The Dead Stars of Population I. *Publications of the Astronomical Society of the Pacific*, 68(403):308, August 1956.

[32] Andrew Dolphin. A New Method to Determine Star Formation Histories of Nearby Galaxies. *New Astronomy*, 2(5):397–409, November 1997.

[33] D. Dravins, L. Lindegren, S. Madsen, and J. Holmberg. Astrometric Radial Velocities from HIPPARCOS. In R. M. Bonnet, E. Høg, P. L. Bernacca, L. Emiliani, A. Blaauw, C. Turon, J. Kovalevsky, L. Lindegren, H. Hassan, M. Bouffard, B. Strim, D. Heger, M. A. C. Perryman, and L. Woltjer, editors, *Hipparcos - Venice '97*, volume 402 of *ESA Special Publication*, pages 733–738, August 1997.

[34] A. S. Eddington. *The Internal Constitution of the Stars*. Cambridge Science Classics, 1926.

[35] Albert Einstein. *The Meaning of Relativity*. Princeton University Press, 1921.

[36] F. R. Ferraro, F. Fusi Pecci, M. Tosi, and R. Buonanno. A Method for Studying the Star Formation History of Dwarf Irregular Galaxies - I. CCD Photometry of WLM. *Monthly Notices of the Royal Astronomical Society*, 241:433–452, December 1989.

[37] William A. Fowler and F. Hoyle. Nuclear Csmochronology. *Annals of Physics*, 10(2):280–302, June 1960.

[38] A. H. Gabriel, J. Charra, G. Grec, J. M. Robillot, T. Roca Cortés, S. Turck-Chièze, R. Ulrich, S. Basu, F. Baudin, L. Bertello, P. Boumier, M. Charra, J. Christensen-Dalsgaard, M. Decaudin, H. Dzitko, T. Foglizzo, E. Fossat, R. A. García, J. M. Herreros, M. Lazrek, P. L. Pallé, N. Pétrou, C. Renaud, and C. Régulo. Performance and Early Results from the GOLF Instrument Flown on the SOHO Mission. *Solar Physics*, 175(2):207–226, October 1997.

[39] Enrique García-Berro, Santiago Torres, Leand ro G. Althaus, Isabel Renedo, Pablo Lorén-Aguilar, Alejandro H. Córsico, René D. Rohrmann, Maurizio Salaris, and Jordi Isern. A White Dwarf Cooling Age of 8 Gyr for NGC 6791 from Physical Separation Processes. *nature*, 465(7295):194–196, May 2010.

[40] N. Giammichele, P. Bergeron, and P. Dufour. Know Your Neighborhood: A Detailed Model Atmosphere Analysis of Nearby White Dwarfs. *Astrophysical Journal Supplement*, 199(2):29, April 2012.

[41] Raffaele G. Gratton, Flavio Fusi Pecci, Eugenio Carretta, Gisella Clementini, Carlo E. Corsi, and Mario Lattanzi. Ages of Globular Clusters from HIPPARCOS Parallaxes of Local Subdwarfs. *Astrophysical Journal*, 491(2):749–771, Dec 1997.

[42] Genevieve J. Graves and Ricardo P. Schiavon. Measuring Ages and Elemental Abundances from Unresolved Stellar Populations: Fe, Mg, C, N, and Ca. *Astrophysical Journal Supplement*, 177(2):446–464, August 2008.

[43] Thomas F. Greene. The Carbon, Nitrogen, and Oxygen Abundances in Four K Giants. *Astrophysical Journal*, 157:737, August 1969.

[44] Jason Harris and Dennis Zaritsky. A Method for Determining the Star Formation History of a Mixed Stellar Population. *Astrophysical Journal Supplement*, 136(1):25–40, September 2001.

[45] C. B. Haselgrove and F. Hoyle. A Mathematical Discussion of the Problem of Stellar Evolution, with Reference to the Use of an Automatic Digital Computer. *Monthly Notices of the Royal Astronomical Society*, 116:515, January 1956.

[46] C. B. Haselgrove and F. Hoyle. A Preliminary Determination of the Age of Type II Stars. *Monthly Notices of the Royal Astronomical Society*, 116:527, January 1956.

[47] Stephen Hawking. Gravitationally Collapsed Objects of Very Low Mass. *Monthly Notices of the Royal Astronomical Society*, 152:75, January 1971.

[48] L. G. Henyey, J. E. Forbes, and N. L. Gould. A New Method of Automatic Computation of Stellar Evolution. *Astrophysical Journal*, 139:306, January 1964.

[49] L. G. Henyey, L. Wilets, K. H. Böhm, R. Lelevier, and R. D. Levee. A Method for Automatic Computation of Stellar Evolution. *Astrophysical Journal*, 129:628, May 1959.

[50] Ejnar Hertzsprung. Über die Sterne der Unterabteilungen c und ac nach der Spektralklassifikation von Antonia C. Maury. *Astronomische Nachrichten*, 179(24):373, January 1909.

[51] Sebastian L. Hidalgo, Adriano Pietrinferni, Santi Cassisi, Maurizio Salaris, Alessio Mucciarelli, Alessandro Savino, Antonio Aparicio, Victor Silva Aguirre, and Kuldeep Verma. The Updated BaSTI Stellar Evolution Models and Isochrones. I. Solar-scaled Calculations. *Astrophysical Journal*, 856(2):125, April 2018.

[52] G. Hinshaw, D. Larson, E. Komatsu, D. N. Spergel, C. L. Bennett, J. Dunkley, M. R. Nolta, M. Halpern, R. S. Hill, N. Odegard, L. Page, K. M. Smith, J. L. Weiland, B. Gold, N. Jarosik,

A. Kogut, M. Limon, S. S. Meyer, G. S. Tucker, E. Wollack, and E. L. Wright. Nine-year Wilkinson Microwave Anisotropy Probe (WMAP) Observations: Cosmological Parameter Results. *Astrophysical Journal Supplement*, 208:19, October 2013.

[53] Jon A. Holtzman, III Gallagher, John S., Andrew A. Cole, Jeremy R. Mould, Carl J. Grillmair, Gilda E. Ballester, Christopher J. Burrows, John T. Clarke, David Crisp, Robin W. Evans, Richard E. Griffiths, J. Jeff Hester, John G. Hoessel, Paul A. Scowen, Karl R. Stapelfeldt, John T. Trauger, and Alan M. Watson. Observations and Implications of the Star Formation History of the Large Magellanic Cloud. *Astronomical Journal*, 118(5):2262–2279, November 1999.

[54] F. Hoyle. The Ages of Type I and Type II Subgiants. *Monthly Notices of the Royal astronomical Society*, 119:124, January 1959.

[55] F. Hoyle and M. Schwarzschild. On the Evolution of Type II Stars. *Astrophysical Journal Supplement*, 121:776–778, May 1955.

[56] F. Hoyle and M. Schwarzschild. On the Evolution of Type II Stars. *Astrophysical Journal Supplement*, 2:1, June 1955.

[57] Edwin Hubble. A Relation between Distance and Radial Velocity among Extra-Galactic Nebulae. *Proceedings of the National Academy of Science*, 15(3):168–173, Mar 1929.

[58] Jr. Iben, Icko. Stellar Evolution Within and off the Main Sequence. *Annual Review of Astronomy and Astrophysics*, 5:571, January 1967.

[59] Jr. Iben, Icko. Age and Initial Helium Abundance of Stars in the Globular Cluster M15. *Nature*, 220(5163):143–146, October 1968.

[60] Jr. Iben, Icko. *Stellar Evolution Physics, Volume 2: Advanced Evolution of Single Stars*. Cambridge University Press, 2013.

[61] Jr. Iben, Icko and John Faulkner. Possible Models of Horizontal Branch Stars. *Astronomical Journal*, 71:165, April 1966.

[62] J. Isern, R. Mochkovitch, E. García-Berro, and M. Hernanz. The Physics of Crystallizing White Dwarfs. *Astrophysical Journal*, 485(1):308–312, August 1997.

[63] P. A. James, M. Salaris, J. I. Davies, S. Phillipps, and S. Cassisi. Optical/Near-Infrared Colours of Early-Type Galaxies and Constraints on Their Star Formation Histories. *Monthly Notices of the Royal Astronomical Society*, 367(1):339–348, March 2006.

[64] Jr. Kennicutt, Robert C. A Spectrophotometric Atlas of Galaxies. *Astrophysical Journal Supplement*, 79:255, April 1992.

[65] T. W.B. Kibble. The standard model of particle physics. *European Review*, 23(1):36–44, 2015.

[66] C. R. King, G. S. Da Costa, and P. Demarque. The Luminosity Function on the Subgiant Branch of 47 Tucanae : a Comparison of Observation and Theory. *Astrophysical Journal*, 299:674–682, December 1985.

[67] R. Kippenhahn, A. Weigert, and Emmi Hofmeister. Methods for Calculating Stellar Evolution. *Methods in Computational Physics*, 7:129–190, January 1967.

[68] M. Koleva, Ph. Prugniel, A. Bouchard, and Y. Wu. ULySS: A Full Spectrum Fitting Package. *Astronomy & Astrophysics*, 501(3):1269–1279, July 2009.

[69] Pavel Kroupa. On the Variation of the Initial Mass Function. *Monthly Notices of the Royal Astronomical Society*, 322(2):231–246, April 2001.

[70] Robert B. Leighton, Robert W. Noyes, and George W. Simon. Velocity Fields in the Solar Atmosphere. I. Preliminary Report. *Astrophysical Journal*, 135:474, March 1962.

[71] G. Lemaitre. Un Univers Homogene de Masse Constante et de Rayon Croissant Rendant Compte de la Vitesse Radiale des Nebuleuses Extra-galactiques. *Annales de la Societe Scientifique de Bruxelles*, 47:49–59, Jan 1927.

[72] D. Lennarz, D. Altmann, and C. Wiebusch. A Unified Supernova Catalogue. *Astronomy & Astrophysics*, 538:A120, February 2012.

[73] John Maddox. Big Bang not yet Dead but in Decline. *Nature*, 377(6545):99, September 1995.

[74] Marie Martig, Morgan Fouesneau, Hans-Walter Rix, Melissa Ness, Szabolcs Mészáros, D. A. García-Hernández, Marc Pinson-neault, Aldo Serenelli, Victor Silva Aguirre, and Olga Zamora. Red Giant Masses and Ages Derived from Carbon and Nitrogen Abundances. *Monthly Notices of the Royal Astronomical Society*, 456(4):3655–3670, March 2016.

[75] T. Masseron and G. Gilmore. Carbon, Nitrogen and α-Element Abundances Determine the Formation Sequence of the Galactic Thick and Thin Discs. *Monthly Notices of the Royal Astronomical Society*, 453(2):1855–1866, October 2015.

[76] M. Meftah, L. Damé, D. Bolsée, A. Hauchecorne, N. Pereira, D. Sluse, G. Cessateur, A. Irbah, J. Bureau, M. Weber, K. Bram-stedt, T. Hilbig, R. Thiéblemont, M. Marchand , F. Lefèvre, A. Sarkissian, and S. Bekki. SOLAR-ISS: A New Reference Spec-trum Based on SOLAR/SOLSPEC Observations. *Astronomy & Astrophysics*, 611:A1, March 2018.

[77] L. Mestel. On the Theory of White Dwarf Stars. I. The Energy Sources of White Dwarfs. *Monthly Notices of the Royal Astro-nomical Society*, 112:583, January 1952.

[78] Marcelo Miguel Miller Bertolami. New models for the Evolu-tion of Post-Asymptotic Giant Branch Stars and Central Stars of Planetary Nebulae. *Astronomy & Astrophysics*, 588:A25, April 2016.

[79] M. Monelli, C. Gallart, S. L. Hidalgo, A. Aparicio, E. D. Skillman, A. A. Cole, D. R. Weisz, L. Mayer, E. J. Bernard, S. Cassisi, A. E. Dolphin, I. Drozdovsky, and P. B. Stetson. The ACS LCID Project. VI. The Star Formation History of The Tucana dSph and The Relative Ages of the Isolated dSph Galaxies. *Astrophysical Journal*, 722(2):1864–1878, October 2010.

[80] W. W. Morgan and N. U. Mayall. A Spectral Classification of Galaxies. *Publications of the Astronomical Society of the Pacific*, 69(409):291, August 1957.

[81] S. Obi. The Structure of Stellar Model with Double Energy Source and its Application to the Study on Stellar Evolution. *Publications of the Astronomical Society of Japan*, 9:26, January 1957.

[82] P. Ocvirk, C. Pichon, A. Lançon, and E. Thiébaut. STECMAP: STEllar Content from High-Resolution Galactic Spectra via Maximum A Posteriori. *Monthly Notices of the Royal Astronomical Society*, 365(1):46–73, January 2006.

[83] B. Paczyński. Evolution of Single Stars. I. Stellar Evolution from Main Sequence to White Dwarf or Carbon Ignition. *Acta Astronomica*, 20:47, January 1970.

[84] Benjamin Panter, Alan F. Heavens, and Raul Jimenez. Star Formation and Metallicity History of the SDSS Galaxy Survey: Unlocking the Fossil Record. *Monthly Notices of the Royal Astronomical Society*, 343(4):1145–1154, August 2003.

[85] Claire Patterson. Age of Meteorites and the Earth. *Geochimica et Cosmochimica Acta*, 10(4):230–237, October 1956.

[86] R. F. Peletier and M. Balcells. Ages of Galaxies Bulges and Disks From Optical and Near-Infrared Colors. *Astronomical Journal*, 111:2238, June 1996.

[87] A. A. Penzias and R. W. Wilson. A Measurement of Excess Antenna Temperature at 4080 Mc/s. *Astrophysical Journal*, 142:419–421, July 1965.

[88] Adriano Pietrinferni, Santi Cassisi, Maurizio Salaris, and Fiorella Castelli. A Large Stellar Evolution Database for Population Synthesis Studies. I. Scaled Solar Models and Isochrones. *Astrophysical Journal*, 612(1):168–190, September 2004.

[89] Adriano Pietrinferni, Santi Cassisi, Maurizio Salaris, and Fiorella Castelli. A Large Stellar Evolution Database for Population Synthesis Studies. II. Stellar Models and Isochrones for an alpha–enhanced Metal Distribution. *Astrophysical Journal*, 642(2):797–812, May 2006.

[90] Planck Collaboration. Planck 2018 Results. I. Overview and the Cosmological Legacy of Planck. *Astronomy & Astrophysics*, 641:A1, September 2020.

[91] T. H. Puzia, S. E. Zepf, M. Kissler-Patig, M. Hilker, D. Minniti, and P. Goudfrooij. Extragalactic Globular Clusters in the Near-Infrared. II. The Globular Clusters Systems of NGC 3115 and NGC 4365. *Astronomy & Astrophysics*, 391:453–470, August 2002.

[92] Harvey B. Richer, Aaron Dotter, Jarrod Hurley, Jay Anderson, Ivan King, Saul Davis, Gregory G. Fahlman, Brad M. S. Hansen, Jason Kalirai, Nathaniel Paust, R. Michael Rich, and Michael M. Shara. Deep Advanced Camera for Surveys Imaging in the Globular Cluster NGC 6397: the Cluster Color-Magnitude Diagram and Luminosity Function. *Astronomical Journal*, 135(6):2141–2154, June 2008.

[93] Harvey B. Richer, Aaron Dotter, Jarrod Hurley, Jay Anderson, Ivan King, Saul Davis, Gregory G. Fahlman, Brad M. S. Hansen, Jason Kalirai, Nathaniel Paust, R. Michael Rich, and Michael M. Shara. Deep Advanced Camera for Surveys Imaging in the Globular Cluster NGC 6397: the Cluster Color-Magnitude Diagram and Luminosity Function. *Astronomical Journal*, 135(6):2141–2154, June 2008.

[94] Adam G. Riess, Peter E. Nugent, Ronald L. Gilliland, Brian P. Schmidt, John Tonry, Mark Dickinson, Rodger I. Thompson, Tamás Budavári, Stefano Casertano, Aaron S. Evans, Alexei V. Filippenko, Mario Livio, David B. Sanders, Alice E. Shapley, Hyron Spinrad, Charles C. Steidel, Daniel Stern, Jason Surace, and Sylvain Veilleux. The Farthest Known Supernova: Support for an Accelerating Universe and a Glimpse of the Epoch of Deceleration. *Astrophysical Journal*, 560(1):49–71, Oct 2001.

[95] R. T. Rood, E. Carretta, B. Paltrinieri, F. R. Ferraro, F. Fusi Pecci, B. Dorman, A. Chieffi, O. Straniero, and R. Buonanno. The Luminosity Function of M3. *Astrophysical Journal*, 523(2):752–762, October 1999.

[96] H. Rosenberg. Über den Zusammenhang von Helligkeit und Spektraltypus in den Plejaden. *Astronomische Nachrichten*, 186(5):71, October 1910.

[97] Vera C. Rubin and Jr. Ford, W. Kent. Rotation of the Andromeda Nebula from a Spectroscopic Survey of Emission Regions. *Astrophysical Journal*, 159:379, Feb 1970.

[98] Tomás Ruiz-Lara, Carme Gallart, Edouard J. Bernard, and Santi Cassisi. The Recurrent Impact of the Sagittarius dwarf on the Star Formation History of the Milky Way. *Nature Astronomy*, 4:965–973, May 2020.

[99] Henry Norris Russell. Relations Between the Spectra and Other Characteristics of the Stars. *Popular Astronomy*, 22:331–351, June 1914.

[100] Henry Norris Russell. Relations Between the Spectra and other Characteristics of the Stars. II. Brightness and Spectral Class. *Nature*, 93(2323):252–258, May 1914.

[101] Henry Norris Russell. On the Sources of Stellar Energy. *Publications of the astronomical Society of the Pacific*, 31(182):205, July 1919.

[102] M. Salaris, S. Cassisi, A. Pietrinferni, P. M. Kowalski, and J. Isern. A Large Stellar Evolution Database for Population Synthesis Studies. VI. White Dwarf Cooling Sequences. *Astrophysical Journal*, 716(2):1241–1251, June 2010.

[103] M. Salaris, S. Degl'Innocenti, and A. Weiss. The Age of the Oldest Globular Clusters. *Astrophysical Journal*, 479:665–672, April 1997.

[104] M. Salaris and A. Weiss. Chronology of the Halo Globular Cluster System Formation. *Astronomy & Astrophysics*, 327:107–120, November 1997.

[105] Maurizio Salaris and Santi Cassisi. Chemical Element Transport in Stellar Evolution Models. *Royal Society Open Science*, 4(8):170192, August 2017.

[106] Maurizio Salaris, Inmaculada Domínguez, Enrique García-Berro, Margarida Hernanz, Jordi Isern, and Robert Mochkovitch. The Cooling of CO White Dwarfs: Influence of the Internal Chemical

Distribution. *Astrophysical Journal*, 486(1):413–419, September 1997.

[107] Maurizio Salaris, Adriano Pietrinferni, Anna M. Piersimoni, and Santi Cassisi. Post First Dredge-Up [C/N] Ratio as Age Indicator. Theoretical Calibration. *Astronomy & Astrophysics*, 583:A87, November 2015.

[108] Maurizio Salaris, Aldo Serenelli, Achim Weiss, and Marcelo Miller Bertolami. Semi-empirical White Dwarf Initial-Final Mass Relationships: A Thorough Analysis of Systematic Uncertainties Due to Stellar Evolution Models. *Astrophysical Journal*, 692(2):1013–1032, February 2009.

[109] Edwin E. Salpeter. The Luminosity Function and Stellar Evolution. *Astrophysical Journal*, 121:161, January 1955.

[110] P. Sánchez-Blázquez, R. F. Peletier, J. Jiménez-Vicente, N. Cardiel, A. J. Cenarro, J. Falcón-Barroso, J. Gorgas, S. Selam, and A. Vazdekis. Medium-rRsolution Isaac Newton Telescope Library of Empirical Spectra. *Monthly Notices of the Royal Astronomical Society*, 371(2):703–718, September 2006.

[111] A. R. Sandage. The Color-Magnitude Diagram for the Globular Cluster M 3. *Astronomical Journal*, 58:61–75, January 1953.

[112] A. R. Sandage and M. Schwarzschild. Inhomogeneous Stellar Models. II. Models with Exhausted Cores in Gravitational Contraction. *Astrophysical Journal*, 116:463, November 1952.

[113] Allan Sandage and Carla Cacciari. The Absolute Magnitudes of RR Lyrae Stars and the Age of the Galactic Globular Cluster System. *Astrophysical Journal*, 350:645, February 1990.

[114] Eric L. Sandquist. A High Relative Precision Colour-Magnitude Diagram of M67. *Monthly Notices of the Royal Astronomical Society*, 347(1):101–118, January 2004.

[115] Ata Sarajedini and Pierre Demarque. A New Age Diagnostic Applied to the Globular Clusters NGC 288 and NGC 362. *Astrophysical Journal*, 365:219, December 1990.

[116] A. Savino, E. Tolstoy, M. Salaris, M. Monelli, and T. J. L. de Boer. The HST/ACS Star Formation History of the Tucana Dwarf Spheroidal galaxy: Clues from the Horizontal Branch. *Astronomy and Astrophysics*, 630:A116, October 2019.

[117] M. Schwarzschild, R. Howard, and R. Härm. Inhomogeneous Stellar Models. V. A Solar Model with Convective Envelope and Inhomogeneous Interior. *Astrophysical Journal*, 125:233, January 1957.

[118] Leonard Searle, W. L. W. Sargent, and W. G. Bagnuolo. The History of Star Formation and the Colors of Late-Type Galaxies. *Astrophysical Journal*, 179:427–438, January 1973.

[119] John R. Stauffer, Greg Schultz, and J. Davy Kirkpatrick. Keck Spectra of Pleiades Brown Dwarf Candidates and a Precise Determination of the Lithium Depletion Edge in the Pleiades. *Astrophysical Journal Letters*, 499(2):L199–L203, June 1998.

[120] H. C. Thomas. Sternentwicklung VIII. Der Helium-Flash bei einem Stern von 1. 3 Sonnenmassen. *Zeitschrift für Astrophysik*, 67:420, January 1967.

[121] Beatrice M. Tinsley. Evolution of the Stars and Gas in Galaxies. *Astrophysical Journal*, 151:547, February 1968.

[122] Eline Tolstoy and Abhijit Saha. The Interpretation of Color-Magnitude Diagrams through Numerical Simulation and Bayesian Inference. *Astrophysical Journal*, 462:672, May 1996.

[123] Floor van Leeuwen. *Hipparcos, the New Reduction of the Raw Data*, volume 350. Springer, 2007.

[124] Don A. Vandenberg, Michael Bolte, and Peter B. Stetson. Measuring Age Differences among Globular Clusters Having Similar Metallicities: A New Method and First Results. *Astronomical Journal*, 100:445, August 1990.

[125] Samuel C. Vila. Pre-White-Dwarf Evolution. I. *Astrophysical Journal*, 146:437, November 1966.

[126] David M. Wilkinson, Claudia Maraston, Daniel Goddard, Daniel Thomas, and Taniya Parikh. FIREFLY (Fitting IteRativEly For

Likelihood analYsis): A Full Spectral Fitting Code. *Monthly Notices of the Royal Astronomical Society*, 472(4):4297–4326, December 2017.

[127] David B. Wood. Multicolor Photoelectric Photometry of Galaxies. *Astrophysical Journal*, 145:36, July 1966.

[128] F. Zwicky. On the Masses of Nebulae and of Clusters of Nebulae. *Astrophysical Journal*, 86:217, October 1937.

Index

Printed in the United States
by Baker & Taylor Publisher Services

Printed in the United States
by Baker & Taylor Publisher Services